Human-Animal Interactions in the Eighteenth

Internationale Forschungen zur Allgemeinen und Vergleichenden Literaturwissenschaft

VOLUME 207

The titles published in this series are listed at *brill.com/favl*

Human-Animal Interactions in the Eighteenth Century

From Pests and Predators to Pets, Poems and Philosophy

Edited by

Stefanie Stockhorst, Jürgen Overhoff, and Penelope J. Corfield

BRILL

LEIDEN | BOSTON

Funding: Subvention received from Potsdam University to reduce price

Cover illustration: François Gaspard Adam: Diana Bathing. Marble sculpture in the park of Sanssouci Palace, Potsdam (1753) [detail]. Photograph courtesy of Stefanie Stockhorst (2021).

Library of Congress Cataloging-in-Publication Data

Names: Stockhorst, Stefanie, 1974- editor. | Overhoff, Jürgen, 1967- editor. | Corfield, P. J., editor.
Title: Human-animal interactions in the eighteenth century : from pest and predators to pets, poems and philosophy / edited by Stefanie Stockhorst, Jürgen Overhoff, and Penelope J. Corfield.
Description: Leiden ; Boston : Brill, [2022] | Series: Internationale forschungen zur allgemeinen und vergleichenden literaturwissenschaft, 0929-6999 ; Vol. 207 | Includes bibliographical references and indexes.
Identifiers: LCCN 2021051237 (print) | LCCN 2021051238 (ebook) | ISBN 9789004448728 (hardback) | ISBN 9789004495395 (ebook)
Subjects: LCSH: Human-animal relationships–History–18th century. | Animals and civilization. | Human ecology.
Classification: LCC QL85 .H8429 2022 (print) | LCC QL85 (ebook) | DDC 590–dc23/eng/20211028
LC record available at https://lccn.loc.gov/2021051237
LC ebook record available at https://lccn.loc.gov/2021051238

Typeface for the Latin, Greek, and Cyrillic scripts: "Brill". See and download: brill.com/brill-typeface.

ISSN 0929-6999
ISBN 978-90-04-44872-8 (hardback)
ISBN 978-90-04-49539-5 (e-book)

This book is printed on acid-free paper and produced in a sustainable manner.

Contents / Table des matières

Figures

Notes on Contributors / Notes sur les contributeurs

Penelope J. Corfield

is Professor Emeritus of History at Royal Holloway, University of London, Visiting Fellow at Newcastle University UK, and President of the International Society for Eighteenth-Century Studies. Her writings focus upon British social, cultural and urban history in the eighteenth and nineteenth centuries, as well as upon general theories of history. On that big theme, she published *Time and the Shape of History* (Yale University Press, 2007); and she contributed an essay on *Time and the Historians in the Age of Relativity*, in Alexander C. T. Geppert and Till Kössler (ed.), *Obsession der Gegenwart: Zeit im 20. Jahrhundert / Obsession with the Here-and-Now: Concepts of Time in the Twentieth Century* (Göttingen: Vandenhoeck & Ruprecht, 2015). Currently, she has a big book in the publication pipeline (for January 2021) entitled *The Georgians: The Deeds and Misdeeds of Eighteenth-Century Britain.*

Kristin Eichhorn

is Assistant Professor for Modern German Studies at Paderborn University (Germany) and board member of the German Society for Eighteenth-Century Studies (DGEJ) since 2018. She has written her PhD thesis on the German fables of the Enlightenment (*Die Kunst des moralischen Dichtens. Positionen der aufklärerischen Fabeldiskussion im 18. Jahrhundert*; Würzburg: Ergon 2013) and is, along with Lothar van Laak, co-editor of the forthcoming book *Kulturen der Moral* (Hamburg: Meiner 2021) to which she contributed the article *Moralische Ästhetik. Konzepte 'nützlicher' Literatur im 18. Jahrhundert*. She is also founder and editor of the journal *Expressionismus* and she has recently published a book (her post-doctoral lecture qualification, 'Habilitation') on Johannes R. Becher (*Johannes R. Becher und die literarische Moderne. Eine Neubestimmung*; Bielefeld: transcript 2020).

Matthew O. Grenby

is Professor of Eighteenth-Century Studies and Dean of Research and Innovation in the Faculty of Humanities and Social Sciences at Newcastle University UK. He was President of the British Society for Eighteenth-Century Studies (BSECS) from 2015 to 2018, and is currently the BSECS representative to the International Society for Eighteenth-Century Studies. He has published books and essays on children's literature and culture, children and heritage, political fiction, and cheap print. In addition, he is now the Principal Investigator on a project funded by the UK Arts and Humanities Research

Council (AHRC) from 2020–2023 to investigate Eighteenth-Century Political Participation and Electoral Culture (ECPPEC).

Anna-Marie Humbert
is a part-time lecturer at the University of Potsdam and curator at the Literaturhaus Berlin. Her fields of interest include Contemporary German Literature, Ecocriticism, and Cultural and Literary Animal Studies. She has recently co-edited a volume on the work of German writer Daniel Kehlmann (*Daniel Kehlmann und die Gegenwartsliteratur*, Berlin: De Gruyter 2020) and she is currently working on weather phenomena in German drama of the early nineteenth century.

Halima Ouanada
est enseignante-chercheuse de littérature française du XVIIIème siècle à l'Université de Tunis El Manar, Présidente de l'Association Tunisienne des Études sur les Lumières (ATEL) et membre de la Société Voltaire et de la Société Internationale d'Études du XVIIIème Siècle (SIEDS). Elle est auteure de *La Femme chez Voltaire*, Tunis, éd. Arabesque 2011; *Térence un poète comique intemporel*, éd. Finzi, Tunis 2007; (*Re*)*penser le féminin* (sous dir.), *L'Harmattan*, Paris 2020. Elle a, en outre, écrit plusieurs articles sur le XVIIIe siècle.

Jürgen Overhoff
is Professor for the History of Education at the Westfälische Wilhelms-Universität Münster. He is President of the German Society for Eighteenth-Century Studies (since 2018) and Delegate Member of the Executive Committee of the International Society for eighteenth-Century Studies (ISECS) since 2018. He is Director of the Center for German-American Educational History at the University of Münster and the General Editor of the Series *Studies on German-American Educational History*. He has edited a variety of volumes and articles on different aspects of the Age of Enlightenment. He is also one of the editors of the bilingual (French-German) Potsdam standard edition of the works of the Prussian King Frederick the Great (*Werke des Philosophen von Sanssouci / Œuvres du Philosophe de Sans-Souci; Friedrich der Große/Fréderic le Grand*).

Kendra Packham
is a Research Associate at Newcastle University UK, where she is part of the project team for the AHRC-funded digital humanities project, Eighteenth-Century Political Participation and Electoral Culture (ECPPEC). She specializes in the relations between literature and political culture in the early

modern period and the long eighteenth century. She is writing a book that will show how a wide range of literature reflected, and helped to shape, an active, far-reaching culture of elections and electioneering between the seventeenth century and the 1832 Reform Act. She has an overlapping interest in representations of, and writings by, labouring-class individuals, and she is completing a book on literature and different discourses about 'the Catholic' during the Restoration and the eighteenth century. Her publications include *Literature and the Culture of Elections and Electioneering in Eighteenth-Century England*, in *The Review of English Studies* (2020).

Antônio Carlos dos Santos
is Full Professor of Ethics and Political Philosophy at the Federal University of Sergipe, Brazil, since 1992. He did a postdoctoral internship at the University of Sherbrooke (2009), Canada, and was a Visiting Professor at the University of Paris I – Sorbonne (2018). He defended his doctoral thesis in philosophy at the University of Paris X – Nanterre, France (2004). His research into political philosophy revolves around seventeenth-century England, notably John Locke, and eighteenth-century France, particularly Montesquieu. He is currently a Delegate Member of the Brazilian Association of Studies of the eighteenth Century (since 2018) on the ISECS Executive Committee. He has published several articles in specialist magazines, book chapters and books, among which the following are particularly notable: *The Two-Way Street: Tolerance and Politics in Montesquieu* (2007), *Denied Politics: Power and Corruption in Montesquieu* (2004) and *The Self in Words* (2018).

Vanessa de Senarclens
est enseignante-chercheuse affiliée à l'Université Humboldt de Berlin où elle a soutenu sa thèse d'habilitation en 2016. Elle est membre du comité exécutif de la société des dix-huitièmistes allemands depuis 2019. Ses travaux portent d'abord sur la littérature française du dix-huitième (Montesquieu, Rousseau, Voltaire), également dans la perspective du transfert culturel. Elle a coédité avec Jürgen Overhoff une édition des œuvres poétiques de Frédéric II. Ses recherches actuelles portent sur l'histoire d'une bibliothèque des Lumières. Sur ce sujet a paru récemment un article retraçant les cheminements d'un volume de Voltaire, depuis sa parution clandestine à Genève jusque dans le dépôt d'une bibliothèque universitaire en Pologne, suite aux déplacements de la frontière germano-polonaise scellés par les accords de Potsdam de 1945: *Verlegt, verwahrt und vergessen. Die Bücher aus den ehemaligen deutschen Bibliotheken in Polen* in: *Merkur. Zeitschrift für europäisches Denken* (octobre 2020).

Stefanie Stockhorst

is Professor for Early Modern German Literature at the University of Potsdam. She was President of the German Society for Eighteenth-Century Studies (2014–18; 2008–10 Vice-President) and has been an Elected Member of the Executive Committee of the International Society for Eighteenth-Century Studies (ISECS) since 2019. She is the general editor of the German Society for Eighteenth-Century Studies' journal *Das achtzehnte Jahrhundert*. In addition to numerous publications on German literary and cultural history (16th–18th c.), she has recently authored a monograph on the cultural history of riding manuals in the Early Modern period: *Ars Equitandi. Eine Kulturgeschichte der Reitlehre in der Frühen Neuzeit* (Hanover: Wehrhahn 2020).

Annika Windahl Pontén

is visitors' coordinator at Uppsala University and Uppsala University Library, Sweden. She holds a PhD in history of ideas and science, and her thesis (2020) is on identity and materiality in the household of Carl Linnaeus. She is an Elected Member of the Executive Committee of ISECS and Treasurer of the Swedish Society for Eighteenth-Century studies. In 2007 she was the project leader for the Linnaeus tercentenary at Uppsala University and she has also worked at a number of museums, including the Linnaeus Museum in Uppsala. She has published articles on dance and music as well as on clothes and textiles in the Linnaeus family. Examples include *En tofs af krusat getragg och fotfodral av matskeväf*, Svenska Linnésällskapets Årsbok (Uppsala 2010) and, together with Roger Tallroth, *'Beständigt menuett och polska'. Om musik och dans i familjen Linné*, Svenska Linnésällskapets årsbok (Uppsala 2017). She was also co-editor of *Låt inte råttor eller mal fördärva: Linnésamlingar i Uppsala. Uppsala universitets jubileumsbok* (Uppsala 2007).

CHAPTER 1

Editorial Introduction

Animals from Pests and Predators to Companions and Cultural Markers

Stefanie Stockhorst, Jürgen Overhoff, and Penelope J. Corfield

Humanity's urgent need to address climate change and thereby to rethink collective attitudes to global biodiversity has injected admirable vitality into historical and cultural studies of human-animal interactions. Powerful inspiration has come not only from historical anthropology, but also from expositions of animal rights and the ethics of animal treatment. Other studies have fruitfully explored motifs and meanings attached to animals in cultural representations. There has also been a lively interest in tracing animal agency in history including the ensuing historiographical challenges[1] – and these days humans are acknowledging, somewhat ruefully, the autonomous activity of a global array of living microbes and viruses too.

Much excellent research has also turned a specific spotlight onto human-animal interactions in the long eighteenth century, also commonly referred to as the 'Age of the Enlightenment'. From the 1970s onwards, Robert Wokler presented seminal articles on the relationship between humans and apes in Rousseau, which gave voice to the notions of 'human animality' and 'perfectible apes'.[2] In 2010 these issues were impressively explored in thematic publications by the *Société française d'étude du XVIIIe siècle*[3] and by the *British Society for Eighteenth-Century Studies*;[4] they indicated the vast array of perspectives to be explored.

The subject area continues to offer rich discoveries for both systematic monographs and specialised case studies. In his recent book, Peter Sahlins, for instance, has explored the influence of animals as cultural icons on politics,

1 Cf. British Journal for the History of Ideas 2 (2017) [special issue: Animal Agents, ed. by Amanda Rees].

2 Now posthumously collected in Robert Wokler: Rousseau, the Age of Enlightenment, and their Legacies. Ed. by Bryan Garsten. Princeton and Woodstock 2012.

3 Cf. Dix-huitième siècle 42 (2010), no. 1 [special issue: L'animal des Lumières, ed. by Jacques Berchtold and Jean-Luc Guichet].

4 Cf. Journal for Eighteenth-Century Studies 33 (2010), no. 4 [special issue: Animals in the Eighteenth Century, ed. by Glynis Ridley].

 | DOI:10.1163/9789004495395_002

society and culture in France at the dawn of the early Enlightenment.[5] Lately, Brycchan Carey has shown how opposition to animal cruelty was closely linked with opposition to slavery and the slave trade.[6] And Avi Lifschitz has surprised and charmed scholars by analysing the sex life of elephants, with reference to the Bible and Enlightenment thought.[7] In her introductory chapter to this volume, Anna-Marie Humbert provides a synoptic overview of other major tendencies in recent eighteenth-century animal studies.

The contributions in the present volume began life as papers for an international workshop entitled *Man's Best Friend? Human-Animal Relations and their Cultural Representations in the 18th Century / Le meilleur ami de l'homme? Les relations humains-animaux et leurs représentations culturelles au XVIIIe siècle*. The programme was originally planned as the scholarly counterpart to the annual meeting of the Executive Committee of the *International Society for Eighteenth-Century Studies* (ISECS), scheduled at the University of Potsdam from 3 to 6 August 2020. Sadly, however, the COVID pandemic forced a change of plan. The Executive Committee was held virtually. At the same time, the Conference presentations were circulated as short reports, which were then open to e-mail comments and criticisms by colleagues, followed by the presenters' responses to the collegial feedback.

These intensive and stimulating exchanges lasted several days. In effect, all contributors got a free consultancy, which they could bear in mind when later expanding their presentations for publication. The resulting volume thus provides not least a physical and durable memento of a set of dynamic intellectual encounters in virtual space.

A further point of note is that the contributors utilise an exceptional range of source materials. Here is documentation based upon (variously): poems, novels, diaries, journals, letters, scientific accounts, philosophical tracts, fables, manuscripts, books, library catalogues, election literature, satirical cartoons, artwork and strikingly unusual objects from material culture, including a horse's skeleton, a 'cheval-machine' (i.e. a mechanical horse), and sundry eighteenth-century volumes which have been devoured by book-worms.

5 Cf. Peter Sahlins: 1668. The Year of the Animal in France. New York 2017.

6 Cf. Brycchan Carey: Abolishing Cruelty: The Concurrent Growth of Anti-Slavery and Animal Welfare Sentiment in British and Colonial Literature. In: Journal for Eighteenth-Century Studies 43 (2020), no. 2, p. 203–220.

7 Avi Lifschitz: The Book of Job and the Sex Life of Elephants: The Limits of Evidential Credibility in Eighteenth-Century Natural History and Biblical Criticism. In: Journal of Modern History 91 (2019), no. 4, p. 739–775.

Specifically, the multi-disciplinary contributions assembled here confront a range of different issues, with a range of original source materials, and with reference to a range of different creatures, from the parrot to the bookworm. The eighteenth-century debates about animals had many ramifications for human understanding not only of the world's profusion of living creatures but also of the distinctive characteristics of humanity. Many theorists stressed particularly the so-called 'anthropological difference'.[8] It distinguished humans not only from animals but also from machines, taking as evidence the intellectual, linguistic, moral and social competencies of the human species.

Meanwhile, traditional taxonomies, like the Aristotelian *scala naturae* ('ladder of nature') or the Biblical order of creation (Genesis 1: 28), which saw 'Man' as the worldly ruler of a hierarchy of animals, were losing their predominance. With regard to the ethical dimension, Jeremy Bentham (1748–1832) made an observation that became seminal for the study of the role of animals in cultural history: "the question is not, Can they reason? nor, Can they talk? but, Can they *suffer*?"[9] Shortly afterwards, his German contemporary Johann Wolfgang von Goethe (1749–1832) would also put the presumed supremacy of the 'rational animal' into perspective. His sarcastic verse exemplifies the spirit of this new way of thinking:

> Er nennt's Vernunft und braucht's allein,
> nur tierischer als jedes Tier zu sein.[10]

Yet cultural attitudes (or mentalities) did not switch suddenly from one register to another. While eighteenth-century scientists were striving to find systematic classifications of animals, people in everyday life coped with the sheer diversity of animalian roles. As the contributions in this volume show, the roles of the 'brute creation' ranged from the negative, as disease-carriers, pests, predators, and metaphors of bestiality; to the neutral, as utilitarian sources of food, raw materials, and horsepower; and through to the positive, as helpmeets, comrades at home and on journeys, exotic show-animals, markers of national identity, metaphors for human characteristics, subjects for scientific enquiry,

---

8 See, e.g., Markus Wild: Anthropologische Differenz. In: Tiere. Kulturwissenschaftliches Handbuch. Ed. by Roland Borgards. Stuttgart 2016, p. 47–59, esp. p. 50–53.

9 Jeremy Bentham: An Introduction to the Principles of Morals and Legislation [1780]. New York 2007, p. 311, emphasis in the original.

10 Johann Wolfgang von Goethe: Sämtliche Werke, Briefe, Tagebücher und Gespräche. Vol. I, 7/1: Faust. Texte. Ed. by Albrecht Schöne. Frankfurt/Main 1994, p. 16 (v. 285–286) ('He calls it reason, using it alone to be far beastlier than any beast'.).

economic contributors, political campaign symbols, material for sayings and songs, and impulses to literary and intellectual creativity. The list is impressive.

Collectively, these intricate analyses of so many different human-animal interactions make a powerful contribution to animal history, which is clearly a branch of the subject that has come to stay.

Lastly, it is a pleasure to express hearty thanks to Masja Horn at Brill for encouragement and support, and to our anonymous peer reviewers for valuable advice throughout this project. In addition, we extend grateful acknowledgement to Vanessa de Senarclens for her careful reading of the French manuscripts before publication. Warm thanks also go to Lena Marie Bombowsky as well as to Sebastian Lange and Luise Seifert, for invaluable assistance as fact-checkers, footnote editors, and proof-readers. The English 'Introduction' was impeccably translated into French by Isabelle Lachance. And our deep appreciation goes to all contributors to this volume, for their admirable mix of scholarly expertise and collegial cooperation.

CHAPTER 2

Introduction des éditeurs

Les animaux : de ravageurs et de prédateurs à des compagnons et des repères culturels

Stefanie Stockhorst, Jürgen Overhoff et Penelope J. Corfield

L'urgence de faire face aux changements climatiques et de repenser ainsi les comportements humains à l'égard de la biodiversité a insufflé une admirable vitalité aux études historiques et culturelles portant sur les interactions entre l'homme et l'animal. Ces études sont d'abord fortement influencées par l'anthropologie historique et par les plaidoyers pour les droits des animaux qui témoignent des avancées éthiques au sujet de leur traitement. D'autres études ont aussi récemment exploré de manière convaincante les motifs et les significations attachés aux animaux dans les représentations culturelles. Elles se sont intéressées à retracer le rôle des animaux dans l'histoire et les défis historiographiques qu'ils posent[1]. A l'heure actuelle, on prend la mesure – non sans dépit – du caractère autonome des activités d'un ensemble de microbes et de virus.

D'excellentes recherches ont également mis l'accent sur les interactions entre l'homme et l'animal au cours du long XVIIIe siècle, communément appelé le 'siècle des Lumières'. À partir des années 1970, Robert Wokler a présenté des articles fondamentaux sur la relation entre les hommes et les singes dans l'œuvre de Rousseau, qui ont permis de définir les notions d''animalité humaine' et de 'singes perfectibles'[2]. En 2010, cette question a été explorée de manière impressionnante dans des publications thématiques à l'initiative de la *Société française d'étude du XVIIIe siècle*[3] et de la *British Society for Eighteenth-Century Studies*[4]; elles signalaient le vaste champ des perspectives à explorer.

1 Voir British Journal for the History of Ideas 2 (2017) [numéro special: Animal Agents, ed. par Amanda Rees].

2 Maintenant réunis dans la collection posthume Robert Wokler: Rousseau, the Age of Enlightenment, and their Legacies. Ed. par Bryan Garsten. Princeton et Woodstock 2012.

3 Voir Dix-huitième siècle 42 (2010), no. 1 [numéro spécial: L'animal des Lumières, ed. par Jacques Berchtold et Jean-Luc Guichet].

4 Voir Journal for Eighteenth-Century Studies 33 (2010), no. 4 [numéro spécial: Animals in the Eighteenth Century, ed. par Glynis Ridley].

 | DOI:10.1163/9789004495395_003

Le sujet continue à offrir de riches découvertes pour des monographies systématiques et des études de cas spécialisées. Dans son récent ouvrage, Peter Sahlins, par exemple, a exploré l'influence des animaux en tant qu'icônes culturelles et politiques dans la France du début du siècle des Lumières[5]. Récemment, Brycchan Carey a montré la manière dont sont étroitement liés le discours d'opposition à la cruauté envers les animaux et celui condamnant l'esclavage et le commerce des esclaves[6]. Et Avi Lifschitz a surpris et aussi fasciné les chercheurs en analysant la vie sexuelle des éléphants, mettant à profit tant les références à la Bible que celles aux auteurs des Lumières[7]. Dans le chapitre introductif du présent volume, Anna-Marie Humbert propose, de son côté, une vue d'ensemble d'autres tendances importantes des études dix-huitièmistes portant sur les animaux.

Les contributions réunies ici ont d'abord eu la forme des communications qui devaient être présentées dans le cadre de l'atelier international intitulé: *Man's Best Friend? Human-Animal Relations and their Cultural Representations in the 18th Century / Le meilleur ami de l'homme? Les relations humains-animaux et leurs représentations culturelles au XVIIIe siècle.* Le programme représentait le pendant scientifique de la réunion annuelle du comité de direction de la Société internationale d'études du XVIIIe siècle (SIEDS), qui devait se tenir à l'Université de Potsdam du 3 au 6 août 2020. Malheureusement, en raison de la pandémie de la COVID, cette réunion a eu lieu uniquement de manière digitale. Les communications ont été diffusées sous forme de brefs résumés à tous les participants qui ont pu réagir et commenter les travaux des autres participants. Il y eut enfin, en dernière étape, la possibilité de réagir et de répondre aux critiques et aux suggestions, ce qui permit un réel échange entre les participants empreint d'un esprit de collégialité.

Ces discussions stimulantes se sont déroulées au cours de plusieurs jours, permettant aux auteurs de bénéficier des conseils offerts et de les intégrer au moment de rédiger les contributions qui composent le présent volume. Ce dernier, issu d'une série de rencontres intellectuelles dynamiques mais virtuelles, offre ainsi un souvenir tangible et durable de cette expérience.

---

5 Voir Peter Sahlins: 1668. The Year of the Animal in France. New York 2017.

6 Voir Brycchan Carey: Abolishing Cruelty: The Concurrent Growth of Anti-Slavery and Animal Welfare Sentiment in British and Colonial Literature. In: Journal for Eighteenth-Century Studies 43 (2020), no. 2, p. 203–220.

7 Avi Lifschitz: The Book of Job and the Sex Life of Elephants: The Limits of Evidential Credibility in Eighteenth-Century Natural History and Biblical Criticism. In: Journal of Modern History 91 (2019), no. 4, p. 739–775.

On notera, en outre, que les contributions ici réunies s'appuient sur un échantillon exceptionnel de sources. Il couvre des poèmes, des romans, des journaux intimes, des revues, des lettres, des comptes rendus scientifiques, des traités philosophiques, des fables, des manuscrits et des imprimés, et aussi des catalogues de bibliothèques, des documents électoraux, des dessins satiriques, des œuvres d'art, des objets étonnamment inhabituels de la culture matérielle – notamment le squelette d'un cheval ainsi qu'un engin appelé le 'cheval-machine' – et divers volumes du XVIIIe siècle perforés par les vers à livre.

Plus précisément, les contributions pluridisciplinaires réunies ici abordent une série de questions diversifiées à partir de sources originales, portant elles-mêmes sur un ensemble de créatures tout aussi hétéroclites allant du perroquet au ver à livre. Au dix-huitième siècle, les nombreuses ramifications des débats sur les animaux ont marqué non seulement la représentation humaine de la profusion des créatures vivantes, mais aussi la conception des caractéristiques de l'être humain. Brandissant les compétences intellectuelles, linguistiques, morales et sociales de l'espèce humaine, de nombreux théoriciens ont alors mis l'accent sur la soi-disant 'différence anthropologique' qui distinguerait non seulement les humains des animaux, mais aussi des machines[8].

Entre-temps, les taxonomies traditionnelles, telles que la *scala naturæ* ('l'échelle de la nature') aristotélicienne ou l'ordre biblique de la Création (Genèse 1, 28), qui concevaient l''Homme' comme le maître incontesté de la hiérarchie animale ont perdu leur prédominance. En ce qui concerne la dimension éthique, l'observation de Jeremy Bentham (1748–1832) a fait école dans l'étude du rôle des animaux dans l'histoire culturelle lorsqu'il remarquait: « La question n'est pas de savoir s'ils peuvent raisonner, ni de savoir s'ils peuvent parler, mais de savoir s'ils peuvent *souffrir* »[9]. À la même époque, l'Allemand Johann Wolfgang von Goethe (1749–1832) relativisait aussi la suprématie présumée de l''animal rationnel' dans des vers sarcastiques illustrant ce nouveau mode de pensée :

> Er nennt's Vernunft und braucht's allein,
> nur tierischer als jedes Tier zu sein[10].

---

8 Voir, par exemple, Markus Wild: Anthropologische Differenz. In: Tiere. Kulturwissenschaftliches Handbuch. Ed. par Roland Borgards. Stuttgart 2016, p. 47–59, esp. p. 50–53.

9 Jeremy Bentham: An Introduction to the Principles of Morals and Legislation [1780]. New York 2007, p. 311 (italiques dans l'original).

10 Johann Wolfgang von Goethe: Sämtliche Werke, Briefe, Tagebücher und Gespräche. Vol. I, 7/1: Faust. Texte. Ed. par Albrecht Schöne. Frankfurt/Main 1994, p. 16 (v. 285–286) ('Il l'appelle raison, et n'a besoin d'elle que pour être plus bestial que n'importe quelle bête'.).

Pourtant, les attitudes culturelles (ou les mentalités) ne changèrent pas si soudainement. Tandis que les scientifiques du XVIIIe siècle s'efforçaient de développer des classifications systématiques des animaux, au quotidien les gens s'accommodaient de la diversité même des rôles attribués aux 'bêtes brutes'. Comme le montrent les textes rassemblés ici, ces rôles oscillaient d'un pôle de connotations négatives – les animaux comme porteurs de maladies, créatures nuisibles, prédateurs et autres incarnations métaphoriques de la bestialité – à un pôle de connotations positives désignant des compagnons de route et de voyage, des acteurs d'un spectacle exotique, des emblèmes d'identité nationale, des incarnations métaphoriques de caractéristiques humaines, des objets de recherches scientifiques des acteurs économiques, des symboles politiques, des matière à proverbes et chansons, des figures de l'élan créateur littéraire et intellectuel. La liste est impressionnante. On trouve aussi à l'intermédiaire entre ces deux pôles une position plus neutre selon laquelle les animaux sont simplement sources de nourriture, de matières premières et d'énergie.

Ensemble, ces analyses complexes portant sur autant d'interactions variées entre l'homme et l'animal proposent une contribution importante à l'histoire des animaux, un domaine de recherche qui s'impose désormais.

Nous souhaitons finalement remercier chaleureusement Masja Horn de Brill pour ses encouragements et son soutien, ainsi que, pour les précieux conseils qu'elle a émis tout au long de ce projet, mais aussi nos collègues ayant procédé à l'évaluation anonyme des contributions réunies dans ce volume. En outre, nous exprimons notre reconnaissance à Vanessa de Senarclens pour sa lecture attentive des textes en français. Nous remercions également chaleureusement Lena Marie Bombowsky ainsi que Sebastian Lange et Luise Seifert pour leur aide précieuse dans la vérification des faits historiques, l'édition des notes et la correction des épreuves. Isabelle Lachance a impeccablement traduit la présente introduction. Et finalement nous tenons à exprimer notre profonde gratitude à tous les auteurs tant pour leur expertise scientifique que pour leur coopération collégiale dont témoignent leurs admirables contributions.

CHAPTER 3

Human–Animal Relations in the Eighteenth Century: Insights from Current Fields of Research

Anna-Marie Humbert

Résumé

Cette contribution traite du lien productif entre les études sur les animaux et celles portant sur le XVIIIe siècle. Il propose d'abord un survol sur les recherches actuelles sur les animaux et poursuit en s'intéressant, plus spécifiquement, à la relation complexe et tendue entre l'homme et l'animal à l'époque des Lumières. L'article montre qu'une compréhension de cette époque dans son rapport aux animaux est non seulement utile, mais aussi nécessaire pour mieux saisir la nature de la relation actuelle entre l'homme et l'animal.

Abstract

This contribution highlights the productive connections between research in the fields of Cultural Animal Studies and of Eighteenth-Century Studies. The first step, in investigating these linkages, is to examine the current state of Animal Studies. And the next step is to dovetail its concerns with the big topics and questions that arise from research into the complex and tense human-animal relationships in the eighteenth century. The outcome is instructive. It shows that studying the century of the Enlightenment is not only rewarding in itself but is actually necessary as a basis for understanding twenty-first-century attitudes to human-animal relationships.

1 Introduction

In a February 2020 issue of the *New York Times*, the Norwegian novelist Karl Ove Knausgård published a long-read portrait of the famous German artist Anselm Kiefer. Their recurring meetings took the two to Kiefer's home town of Donaueschingen, where they visited a local museum and encountered an extensive collection of eighteenth-century animals.

 | DOI:10.1163/9789004495395_004

> I didn't know what was waiting for us inside the museum, but what I laid eyes on as we entered the rooms on the ground floor was nonetheless unexpected: snarling on the floor, with limbs spread wide, was a stuffed crocodile. In the corner behind it stood a huge bear on its hind legs with gaping jaw and forepaws extended; in the cabinets at the back, stuffed birds stood wing to wing; and in the middle of the floor there was a glass case with a leopard in it. [...] It was a cabinet of curiosities, a collection of creatures from far and near, from a time when the world had not yet been fully mapped and when science had not yet subjugated everything. The animals were unusually well preserved; all of them looked alive where they were standing, as if they had been caught yesterday and not in the 18th-century.[1]

Animals from the eighteenth century can be found in various discourses and contexts. They introduce posterity to an era, in which the "world had not yet been fully mapped", as Knausgård put it, but was well underway to be so. They appear, for example, as (exotic) hunting prey and thus also as objects for display, as noted in the above-mentioned quotation. They reveal thereby something about cultural-historical practices such as hunting and taxidermy; and thus about ways of gaining social status and prestige, so indicating the aesthetic preferences of this particular historical period. Furthermore, non-human animals also appear as "textual animals",[2] that is, as narrative figures or anthropomorphised protagonists within traditional fables and tales. In this way, they illustrate the popular literary approaches of their era, whilst also – and above all – demonstrating something about the human self-understanding of 'man' within the historical world order.

In the light of such considerations, this contribution aims to illustrate briefly the wide range of concepts and ideas which then addressed the intense relationships between humans and animals. Evidence is taken initially from the lively interdisciplinary field of Cultural Animal Studies. Focusing on the eighteenth century means exploring not only the past but also the present, since it was then that the moral-philosophical and ethical questions, which to this day shape the

1 Karl Ove Knausgård: Into the Black Forest with the Greatest Living Artist. In: The New York Times, 12 February 2020; online: https://www.nytimes.com/2020/02/12/magazine/anselm-kiefer-art.html [16.12.2020].

2 In the original: "Texttiere". Roland Borgards: Tiere und Literatur. In: idem. (ed.): Tiere. Ein kulturwissenschaftliches Handbuch. Stuttgart 2016, p. 225–244; Roland Borgards: Tiere in der Literatur. Eine methodische Standortbestimmung. In: Herwig Grimm and Carola Otterstedt (ed.): Das Tier an sich. Disziplinenübergreifende Perspektiven für neue Wege im wissenschaftsbasierten Tierschutz. Göttingen 2012, p. 87–118. Unless otherwise noted, all translations are mine.

discourse on animal-human relationships, first became hotly debated. Not only did eighteenth-century thinkers debate intensively the so-called 'anthropological difference' as a criterion for distinguishing non-human animals from human animals, but the question of their basic kinship was also brought into focus. After all, it was the century when Jeremy Bentham (1748–1832) formulated his famous and momentous formula: "Can they [sc. non-human animals] *suffer?*"[3]

Specifically, the discussion that follows provides: firstly a brief overview of the origins and current state of the highly-differentiated field of Cultural Animal Studies; and secondly: an analysis of research topics and approaches that are shared by Animal Studies and Eighteenth-Century Studies. Given the vigour and diversity of both fields, this overview is inevitably selective. Yet it highlights many key insights that inform these broad fields of research.

2 Animal Studies, Human–Animal Studies, Cultural Animal Studies

Climate change, factory farming, the mass extinction of species: those are just three much discussed buzzwords in the twenty-first century. To convey the momentous impact that humans are having on the planet, the term Anthropocene has been used for the last twenty years, indicating the current geological era in which human activities have become predominant factors influencing the global environment and climate. The Humanities in particular, including Cultural Studies, have productively adopted the term, which was introduced and popularised by the Dutch, Nobel Prize-winning meteorologist Paul J. Crutzen in 2000. In the past decade, numerous concepts and theories have emerged under the umbrella term Anthropocene.

Current research on the Anthropocene is part of a branch of research that focuses on humans and their interaction with the natural environment. Developing an interest in non-human animals certainly cannot and should not be justified exclusively with reference to a *Zeitgeist,* sensitised to environmental destruction and animal protection, for the entire history of humanity cannot be meaningfully discussed without involving the non-human species. In order to conduct comprehensive research in the Humanities, the relationships between the human and the non-human world have to be included. Hence, even if the emergence of Animal Studies as a field of research coincides with the rise of the global animal rights movement and has at least partly emerged

3 Jeremy Bentham: An Introduction to the Principles of Morals and Legislation [1780]. New York 2007, p. 311, emphasis in the original.

from it, it is too short-sighted to attribute the growing research interest in non-human animals merely to the political debates of recent decades. Animals are an integral component of anthropological cultural history; the fascination with them has existed since time immemorial.

Collectively, these discourses revolving around the Anthropocene in Cultural Studies and in the Humanities can be assigned primarily to the interdisciplinary research field of Ecocriticism, which is committed to a critical examination of the relationship between humans and the environment. Ecocriticism thus designates an interdisciplinary research field as well as a research paradigm: it focuses an ecologically trained perspective upon cultural artefacts such as texts, images and films.

Much the same point applies to Animal Studies, which emerged at about the same time and which can also be described in terms of these two attributions: they denote a widely-differentiated field of research and a new paradigm, perhaps even a conceptual turn within the Humanities and Cultural Studies. Even if the status of that 'turn' has been discussed for about ten years,[4] a look at the numerous research institutions, networks and publication outlets unquestionably testifies to the infrastructural anchoring of Animal Studies in the current research landscape, especially within the Humanities and Cultural Studies.[5] Animal Studies established itself as a field of research in the USA as early as the 1980s. The basis for the institutionalisation of the field was laid by the Forum Society & Animals, founded in 1981, and the Institute for Animals and Society in Baltimore, founded a few years later (known since 2005 as the Animals & Society Institute).[6] Peter Singer's seminal study *Animal Liberation,*

4 See, for instance, Harriet Ritvo: On the Animal Turn. In: Daedalus. Journal of the American Academy of Arts and Sciences 136 (2007) no. 4, p. 118–122; Ina Bolinski and Stefan Rieger (ed.): Das verdatete Tier. Zum Animal Turn in den Kultur- und Medienwissenschaften. Stuttgart 2019 (Cultural Animal Studies Vol. 5); Friedrich Jaeger (ed.): Menschen und Tiere. Grundlagen und Herausforderungen der Human-Animal Studies. Stuttgart 2020 (Cultural Animal Studies Vol. 9).

5 See, for instance, journals: Society & Animals: The Journal of Human-Animal Studies (1993–); HUMaNIMALIA. A Journal of Human/Animal Interface Studies (2009–); Animal Studies Journal (2012–); Tierstudien (2012–). – Introductions: Margo DeMello: Animals and Society. An Introduction to Human-Animal Studies. New York 2012; Paul Waldau: Animal Studies. An Introduction. Oxford 2013; Gabriele Kompatscher, Reingard Spannring and Karin Schachinger: Human-Animals-Studies. Eine Einführung für Studierende und Lehrende. Münster and New York 2017. – Handbooks: Garry Marvin and Susan McHugh (ed.): The Routledge Handbook of Human-Animal Studies. London 2014; Linda Kalof (ed.): The Oxford Handbook of Animal Studies. New York 2017. – Associations: British Animal Studies Network (2007–) and Australasian and Animal Studies Association (2005–).

6 Cf. Mieke Roscher: Human-Animal Studies. In: Docupedia-Zeitgeschichte; online: http://docupedia.de/zg/roscher_human-animal_studies_v1_de_2012 [07.01.2020].

published as early as 1975, should be mentioned as a pioneer; in addition, the work of Margo DeMello (* 1964) has been decisive for research in Animal Studies.[7]

In a nutshell: Animal Studies is a forty-year-old international multi-disciplinary field of research – with contributions from Philosophy, Literary Studies, History, Biology, Agricultural Sciences, Veterinary Medicine, and Law – which places the animal, long overlooked or at least little considered, at the heart of its research interests. Human-Animal Studies focus on the relationship between animals and humans. DeMello describes it as an "interdisciplinary field that explores the spaces that non-human animals occupy in human social and cultural worlds and the interactions which humans have with them. Central to the field is an exploration of the ways in which animal lives intersect with human societies".[8] This quotation illustrates the basic epistemological assumption of Human-Animal Studies (HAS) that the non-human animal can only ever be studied from the perspective of humans and in comparison with them. This point also raises a further key consideration, much-discussed under the umbrella of Posthumanism, as to how a non-anthropocentric perspective on animals can succeed. Behind this debate lies the question: Can the anthropocentric perspective be discarded or at least deconstructed, as much as possible?

Cultural Animal Studies (CAS), on the other hand, focuses on central research topics, questions, and methods within the related fields of Cultural Studies and the Humanities. In Europe, the animal turn (or at least: animal focus), as imported from the USA and Australia, has probably found its greatest resonance in the Historical Sciences and Philosophy, but it has also been taken up productively in other disciplines, over the past ten to twenty years.[9]

This present volume can accordingly be located in the broad field of Cultural Animal Studies. The field encourages research to be undertaken in a genuinely interdisciplinary way. The focus on the non-human animal opens up the possibilities both of making new findings and of rethinking earlier studies. Indeed, new comparative approaches to re-reading canonical texts hold

7 See Peter Singer: Animal Liberation. A New Ethics for our Treatment of Animals. New York 1975; Margo DeMello: Teaching the Animal. Human Animal Studies Across the Disciplines. New York 2010; eadem, Animal and Society.

8 DeMello, Animals and Society, p. 4.

9 Cf. Roland Borgards: Cultural Animal Studies. In: idem. (ed.): Tiere. Kulturwissenschaftliches Handbuch. Stuttgart 2016, p. 1–6. – See further: Pascal Eitler and Maren Möhring: Eine Tiergeschichte der Moderne: Theoretische Perspektiven. In: Traverse. Zeitschrift für Geschichte. (2008), no. 3, p. 91–105; Erica Fudge: A Left-Handed Blow: Writing the History of Animals. In: Nigel Rothfels (ed.): Representing Animals. Bloomington 2002 (Theories of Contemporary Culture Vol. 26), p. 3–18.

out the prospect of gaining significant new insights. This point does not deny that the field of Cultural Animal Studies has its own genuine research questions and also addresses entirely unknown historical data. However, academic practice – and something similar can be said for Ecocriticism – shows that past researches are often re-read in the context of Cultural Animal Studies. This process should not, however, be seen as a disadvantage, as Cultural Animal Studies contribute in this way to fulfilling the core task of scholarly enquiry: questioning supposed evidence and supplementing existing knowledge with new knowledge.

Earlier contributions to Cultural Animal Studies discussed the animal more as an object of human art and as a target of human action. Today, however, approaches that argue in an anthropomorphizing way – and thus uncritically perpetuate the core difference between human and non-human animals – have become rare. In 1980, John Berger already emphasised in his seminal essay *Why we look at animals*, that the animal has always been at the centre of the world alongside 'man'.[10] Also Paul Shepard's widely acclaimed study, published in 1996, also stressed the cultural co-evolution of humans and animals, highlighting their interconnectedness. Indeed, the title of his work revealed the need for a reversed direction of argument: *The Others: How Animals Made Us Humans*.[11] Twenty years after Shepard, the Dutch primatologist Frans de Waal also points to the necessary reversal of perspective when he asks in his very successful popular study: "Are we smart enough to know how smart animals are?"[12] Moreover, there were much earlier signs of a reversal of a strictly anthropocentric perspective, when in the sixteenth century Michel de Montaigne (1533–1592) asked rhetorically: "When I play with my cat who knows if I am not a pastime to her more than she is to me?"[13]

Throughout the twentieth century, debates about the supposed dichotomy between nature and culture and the alleged 'anthropological difference' were current within sociological and philosophical thought. These days, however, these discussions seem to have been resolved, with a prevalent consensus on the inseparability of both spheres. The abandonment of differentialist arguments testifies to this change – as seen, for example, in the theoretical approaches of writers like Jacques Derrida, Giorgio Agamben, Bruno Latour

10 Cf. John Berger: Why Look at Animals? In: idem: About Looking. New York 1980, p. 1–30.

11 Cf. Paul Shepard: The Others. How Animals Made us Humans. Washington / DC 1996.

12 Cf. Frans de Waal: Are We Smart Enough to Know How Smart Animals Are? New York 2016.

13 In the Original: "[c] Quand je me jouë à ma chatte, qui sçait si elle passe son temps de moy plus que je ne fay d'elle?" (II, 12: 452). Michel de Montaigne: Les Essais. Livre II, ch. XII. Ed. by Pierre Villey and Verdon Louis Saulnier. Paris 2004, p. 452.

and Donna J. Haraway.[14] Those last two authors are particularly pivotal to the current state of Cultural Animal Studies theory. In the work of Latour and Haraway, 'nature' is no longer understood as a sphere opposed to culture. It is rather the idea of the network that is at the centre of their theories. In other words, there are multiple relationships between all agents, as Latour phrases it; or between the 'companion species', as Haraway puts it. This perception raises the central question of agency or, in other words, the efficacy of non-human animals and of things.[15]

Current contributions are increasingly characterised by an aspirational concept of symmetry between the species. This point is also evident in the newly habitual terminology, which no longer speaks of humans and animals, but of non-human and human animals. The concept of *othering,* imported from philosophy and sociology is also used to question the social construction of the animal-human distinction.[16] This constructive development shows how concepts developed in the course of researching marginalised groups in human societies are also being transferred to researching the representation and efficacy of non-human animals.

This practice aims at egalitarian relationships and thus can in part be seen as accusatory, which has also led to generic criticisms of Animal Studies. Critics have accordingly questioned whether scholars in this field maintain an impartial scientific position or rather have shifted to a commitment to animal rights. The commitment to research, allied with a corresponding campaigning attitude, can be found, for example, within the research branch of Critical Animal Studies, which locates its academic work in close proximity to the animal rights movement. Nonetheless, this juxtaposition is by no means true for the entire research field.[17]

---

14 In his remarks on the "anthropological machine", Giorgio Agamben sketches a "zone of indeterminacy" that blurs the boundary between humans and animals; Jacques Derrida problematises the collective-singular "the animal" and shows how ambiguous the constructed boundary to the animal is; Bruno Latour elevates the non-human animal in his network-theory to an agent with agency; Donna Haraway introduces the term "companion species". – Giorgio Agamben: The Open. Man and the Animal. Stanford 2004; Jacques Derrida: The Animal I am Therefore. Transl. by David Willis. New York 2008 [1st ed. in French 2006]; Bruno Latour: Politiques de la nature: Comment faire entrer les sciences en démocratie. Paris 1999; Donna Haraway: The Companion Species Manifesto. Dogs, People and Significant Otherness. Chicago 2003; eadem: When Species Meet. Minneapolis 2008.

15 Cf. Latour, Politiques de la nature; Haraway, When Species Meet.

16 Cf. Kompatscher, Spannring and Schachinger, Human-Animals-Studies, p. 31.

17 See for instance Steven Best: The Rise of Critical Animal Studies. Putting Theory into Action and Animal Liberation into Higher Education. In: Journal for Critical Animal Studies 7 (2009), no. 1, p. 9–52; as well as the *Journal for Critical Animal Studies* (2003–), which "was designed to develop activist consciousness of animal liberation history,

While particular ecological stances, such as opposition to factory farming, to the destruction of natural habitats, and to the extinction of species (as mentioned earlier), are discussed within the conceptual framework of the Anthropocene, it is also apparent that other studies dedicated to a selected animal species, in the context of an overarching Humanities and Cultural Studies research agenda, have recently become popular. This trend can be illustrated by recent studies of bees and bee mortality. Such researches not only illuminate the historically grown interconnection of humans and insects and their rich cultural-historical representation (in, for example, religious poetry, visual art and narrative science) but equally highlight the increasing destruction of the environment and the resource-unfriendly management, perpetrated by humans.[18] Eco-concerns and history can thus fuse together.

3 Human–Animal Relationships in the Eighteenth Century

As is well known, the eighteenth century was a time of discussion and argument about metaphysical, natural and social orders, moral philosophical and ethical conditions, and human duties and freedoms. The non-human animal was also an essential part of these varied discourses: "the eighteenth century was a time not only of rapid expansion of the natural world", as Glynis Ridley put it, "but also of rapidly changing relationships between human and non-human animals".[19] The abundance of topics and contexts concerning human-animal relations is also reflected in the boom in research dedicated to the period.[20] Researchers in Eighteenth-Century Studies have presented various studies that touch on central research questions within the interlinked fields of Human-Animal Studies and Cultural Animal Studies. Two publications call

practice, theory, and politics, while also encouraging Critical Animal Studies (CAS) scholarship"; online: https://journalforcriticalanimalstudies.org/ [17.12.2020].

18 See, for instance, Ralph Dutli: Das Lied vom Honig. Eine Kulturgeschichte der Biene. 3rd ed., Göttingen 2012; Thomas D. Seeley: Honeybee Democracy. Princeton 2010.

19 Glynis Ridley: Introduction. Representing Animals. In: Journal for Eighteenth-Century Studies 33 (2010), no. 4 [special issue: Animals in the Eighteenth Century, ed. by Glynis Ridley], p. 431–436, here p. 432.

20 See, for instance, Nathaniel Wolloch: The Enlightenment's Animals. Changing Conceptions of Animals in the Long Eighteenth Century. Amsterdam 2019; Heather Keenleyside: Animals and Other People. Literary Forms and Living Beings in the Long Eighteenth Century. Philadelphia 2016; Matthew Senior (ed.): A Cultural History of Animals in the Age of Enlightenment. Oxford 2007 (Cultural Histories Series Vol. 4); Frank Palmeri: Humans and Other Animals in Eighteenth-Century British Culture: Representation, Hybridity, Ethics. Aldershot 2006; Aaron Garrett (ed.): Animal Rights and Souls in the Eighteenth Century. Sterling 2000.

for special mention here. Two journals each devoted a special issue to the intricacies of animal-human relationships in the age of Enlightenment: one being the British *Journal for Eighteenth-Century Studies*, the other being the French journal *Dix-huitième siècle*.[21] Scholars contributing to both journals discuss texts focusing upon animal-human relationships, written by both well-known and lesser-known authors. These publications provide an illuminating introduction to the study of animal-human relations in the eighteenth century and simultaneously continue existing research within Cultural Animal Studies.

Above all, the eighteenth century was an era of expanding knowledge and thus also of growing knowledge about non-human animals. As a result of countless research expeditions, the number of known animals doubled during the century.[22] The history of animals in the eighteenth century is therefore also correlated with Europe's expanding colonialism in that era. The increase in knowledge also gave rise to the need for new principles of order. The most effective of these came from Georges-Louis Leclerc de Buffon (1707–1788) and Carl Linnaeus (1707–1778) – and their systems could not be more different. Linnaeus' *Systema naturae* (1735) used tables and lists to create a strict taxonomy of species. Buffon set himself decidedly apart from this in his highly influential eighteenth-century natural history, the *Histoire naturelle* (1749–67), in which he relied on a narrative style.[23] A hundred years later, in the course of the separation of natural science from the fine arts, his seminal work was no longer considered as science but as literature – and it is still the subject of research in Literary Studies today.[24]

Linnaeus' ordering principle of binominal nomenclature, on the other hand, structures to this day, the prevailing scientific approach to the classification of species. As Heinrich Detering points out in his comprehensive study of ecological history, it was remarkable that Linnaeus also included humans in his taxonomy. The genus 'Homo' was assigned to the 'Quadrupedia' – a new kind of inclusion of 'man' into the zoological order.[25] Despite all other differences,

---

21 Cf. Journal for Eighteenth-Century Studies 33 (2010), no. 4 [special issue: Animals in the Eighteenth Century, ed. by Glynis Ridley]; Dix-huitième siècle 42 (2010), no. 1 [special issue: L'animal des Lumières, ed. by Jacques Berchtold et Jean-Luc Guichet].

22 Cf. Ridley, Introduction, p. 432.

23 Cf. Tanja van Hoorn: Naturgeschichte in der ästhetischen Moderne. Max Ernst, Ernst Jünger, Ror Wolf, W. G. Sebald. Göttingen 2016, p. 21–23.

24 Cf. Wolf Lepenies: Das Ende der Naturgeschichte. Wandel kultureller Selbstverständlichkeiten in den Wissenschaften des 18. und 19. Jahrhunderts. Munich and Vienna 1976. p. 150.

25 Cf. Heinrich Detering: Menschen im Weltgarten. Die Entdeckung der Ökologie in der Literatur von Haller bis Humboldt. Göttingen 2020, p. 119–123.

Buffon also grouped animals according to species. However, he enriched this systematisation with detailed narratives of individual animals and maintained a strictly anthropocentric approach, by always focusing on the benefit for humans. This point is stressed by Tanja van Hoorn, in her very illuminating introduction to the function and historical change of key concepts in natural history.[26] For Buffon, it was still 'man' who dominates and completes God's creation.

In addition to Buffon's highly popular natural history storytelling, animals were also used to illustrate complex theoretical issues. Alexander Kling outlines how animal metaphors were repeatedly used in eighteenth-century political thought to discuss models of society. Favourite models were those of the shepherd and his flock of sheep. Thus, for example, in his seminal work *Du contrat social* (1762), Jean-Jacques Rousseau (1712–1778) drew on the image of the shepherd as the pastoral sovereign of the state and the people as a flock of sheep to give a sharp critique of Thomas Hobbes' authoritarian political theory.[27] In a similar vein, popular fables drew heavily upon the symbolic meanings of non-human animals. They tended to be massively anthropomorphised, marking at once their otherness but also their propinquity to humans. In these fables, the non-human animals usually represented prototypical human (mis-) behaviour. In that way, the kindred otherness of the protagonists was needed for the moral education provided by the fables to succeed.[28]

But the non-human animal not only helped to illustrate complex ethical and sociopolitical problems and ideas. It also contributed in a very practical way to the formation of national identities. For instance, the aggregate consumption of meat grew increasingly as, on the one hand, hunting culture shifted and, on the other, meat became more affordable. This shift marked a change in the culture of dining and significantly shaped English identity, as Nicole Maceira Cumming notes: "[B]y the eighteenth century, beef became associated with

26 Tanja van Hoorn, Naturgeschichte, p. 23. – See Georges Louis Leclerc de Buffon: Allgemeine Naturgeschichte. Frankfurt/M. 2008 [*Histoire naturelle, générale et particulière*, 1749–1767], p. 31.

27 Cf. Alexander Kling: Die Tiere der Politischen Theorie. In: Roland Borgards (ed.): Tiere. Kulturwissenschaftliches Handbuch. Stuttgart 2016, p. 97–110, here p. 106–107; Jean-Jacques Rousseau: Du contrat social ou Principes du droit politique. Vom Gesellschaftsvertrag oder Grundsetze des Staatsrechts. Französisch/Deutsch. In cooperation with Eva Pietzcker transl. and ed. by Hans Brockard. Stuttgart 2010, p. 133.

28 See for eighteenth-century German fables in the context of moral education Kristin Eichhorn's contribution in this volume as well as her study: Die Kunst des moralischen Dichtens. Positionen der aufklärerischen Fabelpoetik im 18. Jahrhundert. Würzburg 2013.

Englishness. Meat began to comprise an essential part of an English diet, and, subsequently, this conveyed a sense of national identity".[29]

Thus, the practice of killing animals changed at this time, as did eating habits. Furthermore, in the eighteenth century, the scientific practice of animal experiments also became established.[30] That shift was encouraged by Descartes' theorization of the animal as a non-sentient automaton, technically highly elaborate models of animals were constructed. Jacques de Vaucanson's world-famous mechanical duck was also the result of this machine paradigm. In short, it was a century when human thought oscillated between a detached coldness[31] and an interested, sometimes even loving, attention to animals.[32] The latter can be illustrated, for example, by the growing trend of keeping domestic pets. This development in turn corresponded with the display of social status, as Ingrid Tague shows in her comprehensive study of pets in eighteenth-century Britain. Pets offered a new way to show what it means to be human and what society should look like.[33]

Dogs are a particularly good example to illustrate the many different functions that non-human and domesticated animals fulfilled in the eighteenth century. As an animal that evolutionarily relied on humankind, the dog has always had a special position in cultural history and thus also in Humanities research.[34] In the eighteenth century, dogs appeared in various contexts: As

29 Nicole Maceira Cumming: The relationship between meat consumption and power in late medieval and early modern England. In: Sloth 5 (2019), no. 1; online: https://www.animalsandsociety.org/human-animal-studies/sloth/sloth-volume-5-no-1-winter-2019/the-relationship-between-meat-consumption-and-power-in-late-medieval-and-early-modern-england/ [18.12.2020].

30 See, for instance, Karen A. Raber: Scientific Animals. Reflections on the Laboratory and its Human-Animal Relations, from DBA to Dolly and Beyond. In: Randy Malamud (ed.): A Cultural History of Animals in the Modern Age. Vol. 6. Oxford 2007, p. 119–138; Anita Guerrini: Experimenting with Humans and Animals. From Galen to Animal Rights. Baltimore 2003.

31 See, for instance, on man's cruelty towards the city animals of Paris Laurence Mall: L'animal et la vérité de l'homme social chez Louis-Sébastien Mercier. In: Dix-huitième siècle 42 (2010), no. 1, p. 217–231.

32 A quite particular example of affectionate treatment of or even devotion to an animal is shown by the example of Samuel Johnson, who even bought oysters for his cat. – See Lisa Berglund: Oysters for Hodge, or, Ordering Society, Writing Biography and Feeding the Cat. In: Journal for Eighteenth-Century Studies 33 (2010), no. 4 [special issue: Animals in the Eighteenth Century, ed. by Glynis Ridley], p. 631–645.

33 Cf. Ingrid H. Tague: Animal Companions. Pets and Social Change in Eighteenth-Century Britain. University Park/PA 2015 (Animalibus: Of Animals and Cultures Vol. 5).

34 See Josef H. Reichholf: Der Hund und sein Mensch. Wie der Wolf sich und uns domestizierte. München 2020; Nicole Hoefs and Petra Führmann: Auf Hundepfoten durch die Jahrhunderte. Kulturgeschichten rund um den Hund. Stuttgart 2009.

herding dogs, dogs were an integral part of farming life; they were crucial animal agents for royal hunting; and they functioned as symbols and ornaments for various ruling dynasties.[35] Paintings of aristocrats with dogs at their side were also becoming very popular at this time. But, as mentioned earlier, dogs were increasingly becoming adopted as pets suitable for the masses too. In mid-eighteenth-century Germany, for example, the practice had even become established of reporting missing dogs in the local newspapers.[36]

Additionally, this animal was also becoming a subject of close interest in its own right, as evidenced by the popular paintings of George Stubbs or Jean-Baptiste Oudry. Their artwork still strongly influence today's view of European animals in the eighteenth century.[37] The pictorial representation of individual animals corresponded too with the expanding genre of fictional and non-fictional texts, whose first-person narrators are animals, as Frederike Middelhoff elaborates in her fascinating study *Literarische Autozoographien*. One example was E. T. A. Hoffmann's famous novel *Lebens-Ansichten des Katers Murr* (1819/21) – featuring a tomcat – and that was preceded by other first-person narrating animals, as in *Lebensgeschichte eines Pudels*, a short autobiography of a poodle, published by Joseph Richter in 1808.[38]

The dog thus fulfilled various functions from the farm animal, to the pet, to a pictorial object, to a first-person narrator. Moreover, depending on the context, each dog indicated the corresponding social status of its human counterpart. The broad elevation of the dog from farm animal to broadly accepted pet in the course of the seventeenth and eighteenth centuries demonstrated that the contemporary understanding of the dog as 'man's best friend' and even as a family member has a long history.[39]

35 Cf. Jürgen Overhoff's contribution in this volume.

36 Cf. Aline Steinbrecher: Vermisst! Vermisstenanzeigen für Hunde in den Wöchentlichen Frankfurter Frag- und Anzeigenachrichten des 18. Jahrhunderts. In: Tierstudien 17 (2020) [special issue: Tiere und Emotionen, ed. by. Jessica Ullrich and Marianne Sommer], p. 47–58.

37 See for instance on the pictorial and literary representation of dying animals in the eighteenth century Iris Därmann: Sterbende Tiere. Oudry, Buffon, Rousseau. In: Philipp Stoellger (ed.): Bild und Tod. Anfangsgründe der Bildanthropologie. Tübingen 2016, p. 97–115.

38 Cf. Frederike Middelhoff: Literarische Autozoographien. Figurationen des autobiographischen Tieres (1789–1922). Stuttgart 2020 (Cultural Animal Studies Vol. 7), p. 3. A similar example is Anna Sewell's (in)famous novel *Black Beauty* (1877), subtitled *The Autobiography of a Horse* and also credits the translator: "Translated from the Original Equine".

39 See on the growing interest in animals as pets in the eighteenth century Karen Raber: From Sheep to Meat, from Pets to People. Animal Domestication 1600–1800. In: Matthew Senior (ed.): A Cultural History of Animals in the Age of Enlightenment. Oxford and New York 2007 (A Cultural History of Animals Vol. 4), p. 73–99; Margo DeMello, Animal and Society, p. 84–95.

However, not only animals which have always been culturally strongly linked to humankind, such as dogs and horses, were finding their way into popular experience and artistic representations.[40] There was also a growing fascination with exotic animals, such as talking birds, crocodiles, and rhinos. The most famous late eighteenth-century example was probably the Indian rhinoceros, Clara, which was exhibited all over Europe, thanks to the travelling menageries.[41] These travelling menageries gave way to zoological gardens a few decades later.

The zoo culture subsequently established itself successfully in the nineteenth century.[42] Thereby, the perception of animals changed anew: the exotic was integrated into a familiar environment and the living exhibits, previously visible only in royal gardens, were made accessible to a larger public. In this context, animals can also be read as an expression of an increasing democratisation of the history of nature and knowledge. However, the practice of keeping them captive and exhibiting them also marked their otherness and highlighted humanity's (often cruel) superiority. The travelling menageries, royal gardens, zoological gardens, and later the zoo vividly demonstrated the multifaceted nature of human perceptions of the great variety of non-human animals.

Even the most inconspicuous animals received increased attention in the eighteenth century. In his cycle of poems *Irdisches Vergnügen in Gott* ('Earthly Pleasure in God'), the German poet Barthold Heinrich Brockes created poems in praise of *Die kleine Fliege* ('The Little Fly'), *Der weiße Schmetterling* ('The White Butterfly') and *Die Ameise* ('The Ant') as signs of the purposeful perfection of God's creation, as understood both in nature and in theology.[43] In the eighteenth century, interest in the small, the miniature, found further expression in the growing field of entomology, namely the study of insects.[44]

---

40 See, for instance, on cultural history of horses and riding Ulrich Raulff: Das letzte Jahrhundert der Pferde. Geschichte einer Trennung. Munich 2015; Stefanie Stockhorst: Ars Equitandi. Eine Kulturgeschichte der Reitlehre in der Frühen Neuzeit. Hanover 2020.

41 Jacques Berchtold and Jean-Luc Guichet: Introduction. In: Dix-huitième siècle 42 (2010), no. 1 [special issue: L'animal des Lumières, ed. by Jacques Berchtold et Jean-Luc Guichet], p. 5–18.

42 Cf. Lothar Dittrich and Annelore Rieke-Müller: Unterwegs mit wilden Tieren. Wandermenagerien zwischen Belehrung und Kommerz 1750–1850. Marburg 1999.

43 Cf. Barthold Heinrich Brockes: Irdisches Vergnügen in Gott. Naturlyrik und Lehrdichtung. Selected and ed. by Hans-Georg Kemper. Stuttgart 1999.

44 Cf. Thomas Ruhland: Insekten und die Regeneration des Lebendigen in der Naturgeschichte des 18. Jahrhunderts. In: Forschungsschwerpunkt "Tier – Mensch – Gesellschaft" (ed.): Vielfältig verflochten. Interdisziplinäre Beiträge zur Tier-Mensch-Relationalität. Bielefeld 2017, p. 103–122. – See also Nathalie Vuillemin: Hydres de Lerne et arbres animés: fantasmagories savantes autour du polype. In: Dix-huitième siècle 42 (2010), no. 1 [special issue: L'animal des Lumières, ed. by Jacques Berchtold and Jean-Luc Guichet], p. 321–338.

Researchers like René-Antoine Ferchault de Réaumur (1683–1757) gained influence. Thomas Ruhland points out that insects were "not only at the heart of taxonomic debates in natural history", but also became "crystallisation points, model organisms, in which fundamental philosophical and theological statements about the faculty of life and soul were condensed".[45] He adds that: "Christian Gottlieb Kratzenstein (1723–1795) saw in them the main argument against the mechanistic philosophy of René Descartes (1596–1650) and the proof of the cognitive faculty of animals – including insects – which enabled them to perform arbitrary movements and to exercise memory".[46]

The prolonged dispute about the existence of an animal soul, that flared up in response to Descartes in the seventeenth century, was continued in the eighteenth century.[47] Diderot's sensualist monism can also be mentioned in this context. He argued that all living beings are sentient and thus softened the sharp distinction between humans and non-human animals.[48] Twenty years later, Jeremy Bentham (1748–1832) discussed the rights of animals and is said to have finally ended the controversy about the animal's soul with his utilitarian approach:[49]

> The day *may* come, when the rest of the animal creation may acquire those rights which never could have been withholden from them but by the hand of tyranny. The French have already discovered that the blackness of the skin is no reason why a human being should be abandoned without redress to the caprice of a tormentor. It may come one day to be recognized, that the number of the legs, the pilosity of the skin, or the termination of the os sacrum, are reasons equally insufficient for

45 Ruhland, Insekten und die Regeneration des Lebendigen in der Naturgeschichte des 18. Jahrhunderts, p. 107. – See also the contribution by Vanessa de Senarclens in the present volume.

46 Ibid.

47 Cf. Markus Wild: Die anthropologische Differenz. Der Geist der Tiere in der frühen Neuzeit bei Montaigne, Descartes und Hume. Berlin and New York 2006 (Quellen und Studien zur Philosophie Vol. 74), p. 16–21; Hans Werner Ingensiep: Tierseele und tierethische Argumentationen in der deutschen philosophischen Literatur des 18. Jahrhunderts. In: Internationale Zeitschrift für Geschichte und Ethik der Naturwissenschaften, Technik und Medizin 4 (1996), no. 2, p. 103–118.

48 Cf. Wolfgang Welsch: Menschen und andere Lebewesen. Eine Ontologie jenseits des anthropischen Denkens. In: Friedrich Jaeger (ed.): Menschen und Tiere. Grundlagen und Herausforderungen der Human-Animal Studies. Stuttgart 2020 (Cultural Animal Studies Vol. 9), p. 23–40, here p. 31.

49 Cf. Markus Wild: [Art.] 'Anthropologische Differenz'. In: Roland Borgards (ed.): Tiere. Kulturwissenschaftliches Handbuch. Stuttgart 2016, p. 47–59, here p. 51.

> abandoning a sensitive being to the same fate. What else is it that should trace the insuperable line? Is it the faculty of reason, or, perhaps, the faculty of discourse? But a full-grown horse or dog is beyond comparison a more rational, as well as a more conversable animal, than an infant of a day, or a week, or even a month, old. But suppose the case were otherwise, what would it avail? the question is not, Can they *reason*? nor, Can they *talk*? but, Can they *suffer*? Why should the law refuse its protection to any sensitive being?[50]

What is remarkable about this quotation is not only the much-quoted question "Can they *suffer*?", which is still used in animal ethics today, but also the analogies drawn by Bentham between non-human animals with other discriminated and marginalised groups in human society. As mentioned at the outset, this drawing of analogies is currently being much explored by researchers in Animal Studies. Furthermore, Heike Baranzke refers to the famous Bentham quotation to explain the roots of the radical animal rights movement of the late twentieth century and emphasises that pioneers such as Peter Singer decidedly took up and continued Bentham's argumentation.[51]

Thus current ethical assessments of animals – like trends such as vegetarianism – are rooted here, as highlighted by Peter Singer: "This set of ideas, which largely arose in the eighteenth century, has led to significant changes in the way we treat animals".[52] The debates about moral status extend into present-day research discussions. The ability to act morally continues to be discussed within philosophy and animal ethics; and it is nowadays also linked to debates about theories of cognition.[53]

Probably most significant for current understandings of both human- and non-human animals was the paradigm shift of modernity par excellence: Darwin's theory of evolution, which has redefined the relationship between humans and animals to this day. Again, Darwin's momentous approach can also be traced to roots in the eighteenth century. Darwin was an avid reader of David Hume, and in preparation for *The Origin of Species*

50 Bentham, An Introduction to the Principles of Morals and Legislation, p. 311.

51 Cf. Heike Baranzke: Braucht die Ethik einen *animal turn*? In: Friedrich Jaeger (ed.): Menschen und Tiere. Cultural Animal Studies. In: idem (ed.): Menschen und Tiere. Grundlagen und Herausforderungen der Human-Animal Studies. Stuttgart 2020 (Cultural Animal Studies Vol. 9), p. 193–209, here p. 197.

52 Peter Singer: Foreword. In: Journal for Eighteenth-Century Studies 33 (2010), no. 4 [special issue: Animals in the Eighteenth Century, ed. by Glynis Ridley], p. 427–430, here p. 432.

53 Cf. Evelyn Pluhar: The Nonhuman Roots of Human Morality. In: Klaus Petrus and Markus Wild (ed.): Animal Minds & Animal Ethics. Bielefeld 2013, p. 197–228.

(1859), he also studied Hume's critique of divine design and his remarks on the anthropological difference. Here, Hume prepared the idea that animals could also be moral beings.[54] Only "Hume's rejection of the Anthropological Difference" could, according to Markus Wild, "prepare the naturalisation of man in the succession of the theory of evolution".[55] Once again, it becomes clear that modern understandings of the relationship between human- and non-human animals did not begin with the nineteenth century or later, but must necessarily be discussed with the eighteenth century in mind.

4 Closing Remarks

No question, the aforementioned examples could be continued endlessly: the wealth of eighteenth-century sources is vast. This cursory overview of the diverse and multifaceted questions of human-animal relations shows how productive are the connections between Eighteenth-Century Studies and all forms of Animal Studies. While much of what has led to a comprehensive and differentiated animal ethic in the present day is rooted in the eighteenth century, cultural analysts should be careful not to conceive of this development as a teleological one. The history of human-animal relations was marked by advances and also regressions; many progressive animal ethical positions were formulated in the eighteenth century, but did not find broad public recognition in their own time.[56]

The past 300 years have certainly not been exclusively a good time for the ever-complex nexus of human-animal relationships. The current extinction of species is only the latest testimony to this melancholy fact. The increasing species extinction of the twentieth and twenty-first centuries is one particularly sad hinge that connects the eighteenth century with the present: It was only about 300 years ago that the great wealth of species on this planet was discovered – and today all humans can watch it grow poorer every day.[57]

54 Cf. Wild, [Art.] 'Anthropologische Differenz', here p. 53. – See also Mark Rowlands: Can Animals be Moral? Oxford 2012.

55 In the original: "Erst Humes Zurückweisung der Anthropologischen Differenz" konnte "die Naturalisierung des Menschen in der Nachfolge der Evolutionstheorie vor[bereiten]". (Markus Wild: [Art.] 'Anthropologische Differenz', p. 53.).

56 Cf. Singer, Foreword, p. 429.

57 See Ursula K. Heise: Nach der Natur. Das Artensterben und die moderne Kultur. Frankfurt/ M. 2010.

Compiling a panorama starting from the enthusiastic increase of knowledge of early modernity to the rather disillusioning state of knowledge in the Anthropocene could be a remarkable research desideratum for both Eighteenth-Century Studies and Cultural Animal Studies. The cultural testimonies of this development could also be recorded and related to one another with reference to digital and quantitative methods.[58] To be able to unfold a panorama of schools of thought and the arts, starting from the time when the "world had not yet been fully mapped",[59] as Knausgård specified, up to the present, in which the world is indeed fully mapped, but increasingly emptied of animals, would be an audacious endeavour. Looking back might also bring humans closer to the non-human animals that have already disappeared, "as if they had been caught yesterday and not in the 18th-century".[60]

58 See, for instance, Lars Langer and Manuel Burghadt [et al.]: The Rise and Fall of Biodiversity in Literature: A Comprehensive Quantification of Historical Changes in the Use of Vernacular Labels for Biological Taxa in Western Creative Literature. In: People and Nature 3 (2021), no. 5, p. 1093–1109.

59 Knausgård, Into the Black Forest with the Greatest Living Artist.

60 Ibid.

CHAPTER 4

Of Dogs and Horses: Frederick the Great and His Dearest Animals

Jürgen Overhoff

Résumé

Frédéric le Grand (1712–1786), roi de Prusse, avait un penchant marqué et parfois étrange pour les animaux. Les perroquets et les singes, mais surtout les chevaux et les chiens, ont été ses fidèles compagnons tout au long de sa vie. Sa vénération pour les animaux n'est pas seulement le résultat d'un amour soudain, mais va de pair avec une profonde réflexion philosophique sur les animaux dans la tradition de Locke, de Newton et de Voltaire. En particulier ses poèmes livrent une source peu connue sur le point de vue singulier du roi de Prusse sur les animaux.

Abstract

Frederick the Great (1712–1786), King of Prussia, had a strong and sometimes strange liking for animals. Parrots and monkeys, but especially horses and dogs, were his faithful companions throughout his life. His veneration for animals was not only the result of his personal emotions but also went hand-in-hand with deep reflections upon animal behaviour, in the philosophical tradition of Locke, Newton and Voltaire. Accordingly, Frederick's poems, which have been little studied in this context, are probed here to provide fresh insight into the Prussian king's unique relationship with animals.

1 All the King's Horses and All the King's Dogs: Frederick the Great and His Love for Animals

Frederick the Great, King of Prussia, who was childless and separated from his wife Elisabeth Christine, was surrounded and entertained by domestic animals and a large number of exotic household pets throughout his life. He owned English horses, Italian dogs, African monkeys, an Arabian camel, as well as a huge array of birds from all corners of the world. Significantly, the very first portrait of Frederick, painted by the French artist Antoine Pesne in 1714, when the

 | DOI:10.1163/9789004495395_005

FIGURE 4.1 The dying king. Oil on canvas by Christian Bernhard Rode (1786). Frederick's dog, the delicate Italian greyhound Superbe, is watching the monarch with sympathy. The loyal animal is much closer to the king than his courtiers. The artist Rode knew very well what dogs meant to the *roi-philosophe*.

future monarch was two-years old, depicts the royal child alongside a sprightly leaping dog and an exuberantly colourful parrot. Seven decades later, when the aged king died in his armchair on 17 August 1786 in the magnificent Sanssouci palace in Potsdam, one of his favourite dogs, the greyhound Superbe, attentively sat on a small chair at his right-hand side, watching the old man's decaying face with concern and utmost sympathy. This moving scene was masterfully captured by the director of the Berlin Academy of Arts, Christian Bernhard Rode, in an oil painting finished only weeks after the king had passed away.

Even in death, the Prussian monarch wanted to be near his animals – which is why he decreed in his last will and testament that his body was to be buried on the terrace of Sanssouci next to the eleven greyhounds that he loved most.[1]

From the earliest age, Frederick was truly fond of animals, and he adored some of his darling pets in an almost eccentric fashion. Not all of his family members approved of his – sometimes bizarre – love of animals. While his

1 Testament du roi. In: Frédéric le Grand: Œuvres. Vol. VI. Ed. by Johann David Erdmann Preuß. Berlin 1847, p. 243–244; online: http://friedrich.unitrier.de/de/oeuvres/6/243/text/ [30.09.2020]. There are several last wills of Friedrich, drafted on different occasions, yet all including similar instructions with respect to his burial.

sisters shared his heartfelt sympathy for delicate and dainty dogs, his father Frederick William I, the so-called 'Soldier king', chided Frederick's allegedly 'girlish' love for playful lap-dogs. As a result, the warrior looked down on his eldest son as an effeminate weakling.[2] When Frederick died, his nephew and successor, Frederick William II, could not accept the 'grotesque' idea that the deceased king was to be buried next to the bleached bones of some dogs. So annoyed was the new ruler of Prussia by his predecessor's last will that he bluntly rejected it. Instead, he ordered Frederick's body to be entombed next to his warrior father in the Potsdam Garrison Church. It took two further centuries, and it needed the courageous political initiative of the then German Chancellor Helmut Kohl, for the king to be finally laid to rest on the terrace of the vineyard of Sanssouci in 1991 – next to the tombs of his dogs and in full accordance with his will.[3]

Frederick's sympathetic attitude towards animals figures large in many popular anecdotes that were already told and published by his contemporaries.[4] Today, these stories find their amused readers in the yellow press,[5] as reports about the king's sometimes surprisingly strange love of animals are apt to entertain contemporary readers from all walks of life. Artistic scenes of Frederick and his favourite pets have continuously been part of German souvenir folklore for more than two centuries: China figurines depicting Frederick and his greyhounds are still in demand and sold in the gift sections of the large department stores in Berlin and Potsdam. In popular imagery, the Prussian monarch is always accompanied by his dogs or horses. It is therefore all the more striking that, hitherto, only a handful of academic studies and articles have focused on Frederick's relationship with animals.[6] Apparently, there is a

2 Cf. Johannes Kunisch: Friedrich der Große. Der König und seine Zeit. Munich 2004, p. 24.

3 Cf. Hans Bentzien: Die Heimkehr der Preußenkönige. Berlin 1991.

4 See esp. Friedrich Nicolai: Anekdoten von König Friedrich II. von Preussen, und von einigen Personen, die um Ihn waren. Nebst Berichtigung einiger schon gedrukten Anekdoten. Berlin and Stettin 1788; Anton Friedrich Büsching: Beyträge zu der Lebensgeschichte denkwürdiger Personen, insonderheit gelehrter Männer. Vol. 5. Halle 1788, p. 22–24.

5 Typical examples of this type of report about Frederick and his animals (all published around the 300th anniversary of the king's birth) are: Katharina von der Leyen: Friedrich II., der große Hundefreund. In: Bild.de, 12 July 2012; Christoph Stollowsky: Der Alte Fritz. Das gute Herrchen. In: Der Tagesspiegel, 7 February 2012; Simon Benne: Friedrich der Große war ein Tierfreund. In: Hannoversche Allgemeine, 29 June 2012; Wolfgang Grittner: Der Preußenkönig Friedrich II. und die Tiere. Eine historische Betrachtung. In: Deutsches Tierärzteblatt 3 (2013), p. 316–319.

6 The best – if short – study on Frederick and his relationship to animals is by Prinzessin Sibylle von Preußen and Prinz Friedrich Wilhelm von Preußen: Friedrich der Große. Vom anständigen Umgang mit Tieren. Göttingen 2012.

FIGURE 4.2 Gravestones of Fredrick the Great and his dogs on the terrace of Sanssouci Castle. The names of the dogs are Alcmène, Thisbe, Diana, Phillis, again Thisbe, Bigos, again Diana, Pax, Superbe, Hasenfuß and Amouretto.

certain kind of embarrassment involved, when scholars, especially old-school Prussian historiographers, try to interpret and make sense of the great king's intense occupation with dogs, birds, horses, and apes.

Johannes Kunisch, author of the standard biography of Frederick the Great,[7] mentions the king's love for animals and dogs only in passing and in a very distant manner. Moreover, he completely ignores the moving scene – reported by Mirabeau – when the aged monarch grieved for his beloved greyhound Alcmène, whom he had laid out in his Potsdam library for three days after the dog's death in 1785.[8]

7 Cf. Kunisch, Friedrich der Große.

8 Honoré Gabriel de Riqueti, comte de Mirabeau: Lexikon aller Anstössigkeiten und Prahlereyen, welche in denen zu Berlin in funfzehn Bänden erschienenen sogenannten Schriften Friedrichs des Zweyten vorkommen. Schönfeld 1790, p. 64.

FIGURE 4.3 Fredrick the great and his dogs in his library at Sanssouci. Chromolithograph by Carl Röchling (1890)

2 Voltaire, Newton, and Locke: Frederick's Philosophical Poetry of Animals

Far from being merely anecdotal, reports about Frederick's affectionate care for his animals reveal that the king was not only sentimentally attached to his pets, but that he also thoroughly reflected on the particular quality and essence of human-animal-relations. Time and again, he tried to define the similarities and crucial differences between the lives of men and the existence of beasts. In the eyes of the remarkable Prussian king – who styled himself *roi-philosophe* and, accordingly, conducted a learned correspondence with the leading Enlightenment thinkers of his era – it was a worthy intellectual task to explore attentively the instincts and cognitive faculties of animals. More than in any other of his writings, his poems – written in French and composed throughout his life – contemplate the behaviour and intentions of animals in the most intimate and intense manner. Hitherto, Frederick's poems have been rarely consulted by interpreters of his philosophy, but focusing on them as a unique source of his views on animals yields some unexpected and fascinating results.[9]

9 Frederick the Great considered his sophisticated poetry as a prime example of his philosophical pursuit. This is clearly indicated by the self-chosen title of his most important collection

A central theme of Frederick's poems is the fundamental difference between the opposing principles of the philosophies of René Descartes and Isaac Newton. The Prussian king sided with Newton, once Frederick's distinguished French correspondent and teacher Voltaire published a popular introduction to the famous English physicist's system of thought. On 3 December 1736, Frederick confessed to Voltaire that he eagerly awaited the publication of Voltaire's *Éléments de la Philosophie de Newton*.[10] When the book finally appeared in 1738, the royal philosopher from Prussia became an ardent and life-long admirer of the Englishman. Frederick referred to him in his poems as "le grand Newton",[11] whereas he chided Descartes as someone who would tragically "blind [éblouir]"[12] the minds of men with false notions of the world. In particular, Frederick counted among Descartes' most blatant errors, his cold understanding of animals as mere machines and automata, unable to feel, calculate, or think.[13] Instead, Frederick held in high esteem the alternative view, which Voltaire had outlined as Newton's doctrine: the English natural philosopher stated clearly that animals were quite obviously bestowed by their divine creator with a certain degree of understanding and even ideas ("une mesure d'idées").[14] In that respect, they could be compared with humans.

If animals were more than mere beast-machines, if they could think, at least in a rudimentary way, then they were not senseless beings. Animals had feelings and a sense of compassion on a level similar to that of humans. This view is what Voltaire had learned from Newton and he passed that insight on to all his readers and to Frederick in particular. In Chapter 5 of his *Éléments*

of poems – Frédéric le Grand: Œuvres du Philosophe de Sans-Souci. Au donjon du château. Avec le privilege d'Apollon. 3 vols. Berlin 1750–1762; New and reliable scholarly editions of his poems are: Friedrich der Große – Potsdam Edition [bilingual]. Vol. VII: Werke des Philosophen von Sanssouci. Oden, Episteln, Die Kriegskunst/Œuvres du Philosophe de Sans-Souci. Odes, Épitres, L'art de la guerre. Ed. by Jürgen Overhoff and Vanessa de Senarclens. Berlin 2012; Jürgen Overhoff and Vanessa de Senarclens (ed.): An meinen Geist. Friedrich der Große in seiner Dichtung. Eine Anthologie [French/German]. Paderborn 2011.

10 Cf. Voltaire – Friedrich der Große: Briefwechsel. Ed. by Hans Pleschinski. Munich 2004, p. 30.

11 E.g. Friedrich der Große, Œuvres du Philosophe, p. 172 and p. 256.

12 Ibid., p. 182.

13 Cf. Peter Harrison: Descartes on Animals. In: The Philosophical Quarterly 42 (1992), p. 219–227.

14 Voltaire: Éléments de la philosophie de Newton [1738]. In: The Complete Works of Voltaire. Ed. by Theodore Besterman. Vol. 15. Oxford 1968, p. 222.

de la philosophie de Newton, Voltaire declared: "Il y a surtout dans l'homme une disposition à la compassion aussi généralement répandue que nos autres instincts: Newton avait cultivé ce sentiment d'humanité, et il l'étendait jusqu'aux animaux".[15] Voltaire also found it important to stress that Newton was in this respect in full accordance with John Locke: "Il était fortement convaincu, avec Locke, que Dieu a donné aux animaux (qui semblent n'être que matière) [...] les mêmes sentiments qu'à nous. Il ne pouvait penser que Dieu, qui ne fait rien en vain, eût donné aux bêtes des organes de sentiment afin qu'elles n'eussent point de sentiment".[16] Locke had indeed emphasized that animals had feelings, and that unnecessary cruelty toward them was ethically wrong;[17] Voltaire had already paid homage to these ideas in his thirteenth letter "Sur M. Locke" of his *Lettres anglaises* of 1734.[18]

Frederick remained in accordance with Voltaire, Newton, and Locke throughout his life. Their sympathetic description of animals as sensible beings that were capable of compassion became a cornerstone of his philosophical poetry. Like Voltaire, he held that one should treat animals with the utmost care and respect and that they deserved no suffering or any other form of cruel handling. Thus, it would be "une contradiction bien affreuse à croire que les bêtes sentent, et à les faire souffrir".[19] But, what is more, humans were not only asked to respect animals as emotionally – and, to a degree, cognitively – equal beings, they were also invited to mirror their own human conduct, or misconduct, in the natural behaviour of dogs, horses, apes, or even insects. When seeking to discover the truth about the human condition, it was necessary to spend a lot of time with animals, to share one's life with them and to watch their actions closely, working out the resemblances, contrasts, and divergences between them and humans. This approach is exactly what Frederick did in some of his most important poems.

What the king wished to convey in his often self-ironic, or even sarcastic, odes and epistles was a chastening sense of humans as a most arrogant, presumptuous, and therefore in many respects cold-hearted species, which ridiculously overestimated its intellectual capacities to the detriment of its moral

15 Ibid.

16 Ibid.

17 Locke wanted humans to treat all animals with care and respect, no matter whether "young Birds, Butterflies, and such other poor Animals", "Squirils" or "Dogs"; cf. John Locke: Some Thoughts concerning Education [1693]. Ed. by John W. and Jean S. Yolton. Oxford 1989, § 116, p. 180.

18 Cf. Voltaire: Lettres écrites de Londres sur les anglois et autres sujets. Basel 1734.

19 Voltaire, Éléments, p. 222.

standards or emotional instincts. In his magnificent epistle "Sur la faiblesse de l'esprit humain", he sceptically concluded that man's mind was often as limited ("esprit limité") as the understanding of a small fly or a tiny ant, which is why Frederick scornfully compared man's poor wits with the sheer ignorance of a laughably vainglorious insect ("insecte orgueilleux").[20] Accordingly, feeble reason ("raison si faible")[21] did not ennoble humans, since this imperfect mental faculty was shared by men and women, cats and dogs, horses and mice. Reason was the highly problematic queen of *all* animals ("Reine des animaux").[22]

What distinguished 'man' from other animals was his morally questionable ability to use language and to invent words ("Il invente des mots")[23] in order to lie and deceive his fellow creatures, to outdo them, and to gain advantage over them. Language was often used by proud and selfish humans as a means to betray or to invent a seductive lie ("le seduisant mensonge") in order to nourish deceptive fictions ("[n]ourri[r] de fictions notre crédulité").[24] This interpretation was not at all flattering, and it contradicted the view that humans surpassed animals because of the human capacity to reason and to speak. In his lengthy epistle *À mon esprit*, Frederick contrasted this kind of ill-meant human rhetoric with the simple and straightforward language of animals ("le langage des bêtes").[25] Animals used their particular language of individual sounds and gestures, as he contended, only to express sincere feelings. They did not always live together in peace or in harmony; but, when they quarrelled, they only followed their natural instincts out of sheer necessity. Humans, on the other hand, fought with words and with deeds for entirely different reasons – either to gain glory or to seek power for power's sake.

In his poem "Les deux chiens et l'homme", written in the style of a fable, Frederick makes this conviction abundantly clear. In these well-composed verses, the king describes in rhymes – and with a heavy pedagogical emphasis – two athletic dogs that have not eaten for some time and are looking for food. Suddenly, they discover a tasty bone. They start to fight with each other, because this unexpected delicacy seems to be far too small to satisfy the needs of both of the contenders. The dogs wildly bark at each other, as they are in the fierce grip of strong and irresistible cravings ("irrités par la faim").[26] A passer-by is so outraged by the extremely loud cries of the two competing rivals

20 Friedrich, Œuvres du Philosophe, p. 170.
21 Ibid., p. 176.
22 Ibid.
23 Friedrich, Œuvres du Philosophe, p. 171.
24 Ibid., p. 172.
25 Friedrich, Œuvres du Philosophe, p. 397.
26 Les deux chiens et l'homme. In: Friedrich, An meinen Geist, p. 256.

that he reaches for a stick to discipline them and to end their noise. Beating them harshly and severely ("dure et brutale"),[27] he acts in an uncontrolled rage. At the same time, he verbally insults the two dogs, denouncing them as inferior beings and deriding them as four-legged mongrels ("canaille quadrupède").[28] Almost beaten to death, the two dogs manage to escape at the very last minute. While they flee, one of the humiliated animals barks back at the man from a secure distance. Full of anger, the dog confronts him with a grave reproach: Dogs fight for bones – humans, on the other hand, go to war and kill each other only to satisfy their lust for power.

The poem was composed in 1762 at the end of the Seven Years' War. The author himself had been involved in this terrible military conflict, a war that was waged globally from 1754, onwards, when British and French forces started to attack each other in North America in order to gain control over that vast continent. As an ally of the British, Frederick entered the European theatre of war in 1756.[29] By the end of this global power struggle, Prussia had barely survived – and it had made no significant conquests. When the peace treaties of Paris and Hubertusburg were signed, Frederick conceded that he had no choice but to return to the *status quo ante*. He had to accept the prewar balance of power, thereby realizing that all the bloodshed and all the heavy losses of thousands of soldiers had been in vain. War was a cruel and futile enterprise. Thus, the royal poet Frederick sounds like a self-critical warrior when he states at the end of "Les deux chiens et l'homme" that dogs only fight for their needs, whereas humans are led in their aggressions by folly and pride: "De vrais besoins entre chiens font les guerres, Entre nous c'est l'orgueil et cent chimères".[30]

The philosopher king unmasked the true intentions of belligerent humans with grim veracity and mortal realism, and he admired – and even envied – dogs for their genuine feelings and unfeigned instincts. At times, he was even tempted to swap roles with them. In one of his last poems, *À une chienne*, the then aged and sick king explicitly states that he wishes to be a bitch, a female greyhound, a most delicate being that is gentler and more peaceful than a male dog. Talking to the fictional bitch, he sighs and confesses "Je t'envie, ô bichon".[31] In his intimate correspondence with his elder sister Wilhelmine, Frederick

27 Ibid.

28 Ibid.

29 Cf. Jürgen Overhoff: Friedrich der Große und George Washington. Zwei Wege der Aufklärung. Stuttgart 2011, p. 217.

30 Ibid.

31 À une chienne. In: Friedrich, An meinen Geist, p. 274.

likewise styled himself as a female dog. In one of these letters he said about himself, that he refrained from calling himself a greyhound, only because he was too lazy to crawl on all fours.[32] Both beings, he emphasized, the king and the dog, had the same – and quite imperfect – intellectual capacity.

Perhaps it was just a sign of the king's proud ownership of amiable dogs that the collar of one of Frederick's beautiful greyhounds was engraved with his own name and royal title "Fridericus Rex". But it could also be that the king wanted ironically to indicate that his dogs were his equals. At any rate, he regularly asked eminent artists to portray his greyhounds, and he painstakingly decorated the walls of his castles in Charlottenburg and Sanssouci with a large number of these extraordinary paintings. The most prominent place in Sanssouci was reserved for a large painting depicting Frederick's white greyhound Biche as the lapdog of the goddess Diana. Other paintings represent his best loved female dogs Alcmène, Superbe, and, again, Biche in a highly realistic manner. They sit on red velvet cushions, giving the impression of being the most precious possession of a man who loves them more than crowns and jewels.

3 Sharing Table, Bed, Library, and Excursions: The King's Everyday Life with His Animals

Frederick's deep philosophical reflections on animals and their seemingly artless feelings and instincts had important consequences for the way he shared his life with them. His dogs were given full freedom to roam and play wherever they wished. He often carried one of his rather dwarfish greyhounds under his coat for hours, while the other dogs frolicked in all the rooms of his palaces, playing with small leather balls specifically designed by the king. When he worked with utmost concentration in his library, they were silent and lay on the floor. They were always around him and he never grew tired of their sight. They were even allowed to share his bed with him. Frederick treated his dogs with more respect than he accorded to his courtiers and servants. Anton Friedrich Büsching, director of the prestigious Berlin Gymnasium zum Grauen Kloster and former professor for philosophy at the University of Göttingen, vividly recalled in his detailed biography of Frederick the Great – published only

32 Gustav Berthold Volz (ed.): Friedrich der Große und Wilhelmine von Baireuth. Vol. II. Leipzig 1924, p. 143.

FIGURE 4.4 One of Frederick's greyhounds. The collar is engraved "Fridericus Rex". Oil on canvas. [Anonymous artist] (18th c.).

two years after the king's death – that the monarch gave his dogs "permission to do whatever they wanted".[33]

Büsching's fascinating report reveals every aspect of the astonishingly cosseted lives of the king's greyhounds. While his favourite dog would "sleep at his side all night in bed",[34] the other greyhounds lay dispersed on various sofas. All the beds, sofas, wooden floors, carpets, and curtains in the castles of Charlottenburg and Sanssouci suffered damages from the uncontrolled behaviour of the king's darling pets. The impetuous dogs gave the precious furniture a rough treatment, the whole interior of Frederick's private rooms was "dirty and tattered".[35] The same observation, by the way, was recorded by Johann Wolfgang von Goethe, who visited Sanssouci Palace with the Duke of

33 Anton Friedrich Büsching: Character Friedrichs des zweyten, Königs von Preußen. 2nd ed., Halle 1788, p. 22: "[D]er König erlaubte ihnen alles".

34 Ibid.: "schlief des Nachts bey Ihm im Bette".

35 Ibid.: "beschmutzet und zerrissen".

Saxony-Weimar-Eisenach, Carl August, on 22 May 1778, a day when the king was away.[36] Yet, Frederick cared for his dogs "in the tenderest manner".[37] He fed them personally with the most delicious food when he dined at his festive table and "once they had eaten, they were taken care of by one of his servants, who walked them when the weather was fine, so that they could enjoy fresh air".[38]

When his dogs were sick, Frederick tried to cure them with a self-invented whey-based diet. With regard to his beloved greyhounds, he did not even trust the royal veterinarians and issued a warning that no self-styled doctor of dogs should touch his greyhounds ("kein Hunde-Doctor sol sie nicht angreifen").[39] But Frederick could not perform miracles and keep them healthy at all times and under all circumstances. Dogs were beings as mortal as humans, and they lived far shorter lives. A decade was a lifespan his greyhounds seldom reached. And when they died, the Prussian monarch was inconsolable. After Biche's death, he wrote a moving letter to his sister Wilhelmine, where he expressed his deep sorrow and confessed his unexpected weakness with regard to the loss of only a small animal: "I was ashamed that the death of a dog did affect me so much, but the domestic life I lead, and the loyalty of the poor animal, made it close to my heart. Its sufferings made me so nervous that, frankly speaking, I am sad and depressed".[40]

Frederick's many greyhounds were undoubtedly his favourite animals; their exceptional status at the royal court was unparalleled. But dogs were not the only species with exclusive and intimate access to the king. The monarch also owned a tame monkey named Mimi, whom he always allowed to sit next to him in the library. He often kept the monkey at bay, putting a chain around his leg. Occasionally, however, he gave the animal full freedom. In one instance, the ape used its liberty in a disruptive way, causing havoc in the library by

36 Goethe was more than astonished to see a whole menagerie living in Frederick's marvelous palace, monkeys as well as parrots ("Affen, Papageien"), and the poet was full of disdain when he noticed that the curtains had been tattered by the king's greyhounds ("zerrissene Vorhänge"), cf. Johann Wolfgang von Goethe to Johann Heinrich Merck, 5 August 1778. In: Johann Wolfgang von Goethe: Briefe. Hamburger Ausgabe. Vol. 1. Ed. by Karl Robert Mandelkow. Munich 1976, p. 253.

37 Büsching, Character, p. 22: "aufs zärtlichste".

38 Ibid.: "überhaupt aber wurden die Hunde von einem Bedienten versorget, der sie auch nach ihrer Mahlzeit bey guter Witterung spazieren führete, damit sie der frischen Luft genießen konnten".

39 Frederick the Great to Michael Gabriel Fredersdorf (1751 or 1752). In: Frédéric le Grand: Œuvres. Vol. XXVII. Ed. by Johann D. E. Preuss. Berlin 1856, p. 312.

40 Volz, Friedrich der Große, p. 237.

throwing books from the shelves – and worse. In a letter to his friend Ulrich Friedrich von Suhm, the king described with black humour the chaotic situation: Cheeky Mimi, his otherwise loyal comrade ("fidèle compagne"), grabbed a book, "holds it into a candle, puts it to flames and applauds when seeing it burning".[41]

Frederick, one of the great bibliophiles of his age, should have been devastated by the loss of a valuable book. He understood, however, that the monkey was not a mean creature, but was following its playful instincts and heedless impulses. So the king did not punish the animal for its inappropriate behaviour. Instead, he used the occasion to blame – once again – the imperfections of human reason. The burned book had been the sixth (and enlarged) edition of the philosopher Christian Wolff's famous treatise *Vernünfftige Gedancken von Gott, der Welt und der Seele des Menschen* (1st ed. 1720), his so-called 'Deutsche Metaphysik' ('German Metaphysics'). Frederick revered Wolff as one of the leading philosophers of the Enlightenment, but he had to admit that his arguments were not always easy to follow. Wolff's literary style often obstructed his philosophical message, as he wrote in a rather complex way. The king therefore ironically assumed that the clever "monkey had wanted to study the Metaphysics, but put it into the flame, because he simply could not understand it".[42] Or perhaps the ape was angry with Wolff, because the philosopher had ranked humans above animals, thereby discrediting the monkey species.[43]

---

41 Frederick the Great to Ulrich Friedrich von Suhm, 22 January 1737. In: Frédéric le Grand: Œuvres. Ed. by Johann D. E. Preuss. Vol. XVI. Berlin 1850, p. 312: "l'allume à la chandelle, et s'applaudit de la voir brûler".

42 Ibid.: "le singe avait voulu étudier la *Métaphysique*, et que, ne l'ayant pu comprendre, il l'avait brûlée".

43 Cf. ibid., p. 313. – Frederick highly regarded Wolff as a philosopher of religious tolerance, and he offered him a professorship at the University of Halle, where Wolff instructed generations of students from 1740 until the year of his death in 1754; cf. Jürgen Overhoff: Christian Wolffs brisante Vorliebe für die altchinesische Philosophie. Eine Erkundung der Grenzen der Toleranz im Halleschen Pietismus und in der preußischen Monarchie (1711–1723). In: Holger Zaunstöck, Brigitte Klosterberg, Christian Soboth and Benjamin Marschke (ed.): Hallesches Waisenhaus und Berliner Hof. Beiträge zum Verhältnis von Pietismus und Preußen. Halle 2017, S. 137–152. – Yet, the king did not agree with Wolff's views on animals as published in his *Vernünfftige Gedancken von Gott, der Welt und der Seele des Menschen, auch allen Dingen überhaupt*. Halle 1720. – For Wolff, animals were entirely irrational creatures. In chapter 5, § 870, p. 478, of this influential book, he took the example of a dog – brutally beaten by a man – in order to explain that animals could feel pain, but were not able to reflect upon the treatment they had suffered by humans. Frederick's poem *Les deux chiens et l'homme*, as interpreted above, tells a different story.

Frederick always protected his animals against false accusations, and he sought to treat them as mildly and gently as possible. This approach was particularly the case with regard to his horses. In an age where horses were primarily used for working purposes – in peace and also in wartime, when thousands of them were terribly wounded or killed in horrible slaughter caused by the hateful actions of men[44] – the king treated many of these animals as loyal companions for leisurely activities. Whereas they were normally used as carriers of all sorts – or to pull carts or carriages, regularly exposed to lashes by whips or rods – the Prussian monarch forbade any rough treatment of his horses. He himself would never use the whip and he also abstained from wearing riding boots with spurs. He simply did not want to hurt his horses by repeatedly kicking them in their unprotected flanks when sitting on horseback. All contemporary paintings of the king therefore correctly represent him without spurs.

The Prussian monarch owned up to 4,000 horses at a time. Most of them were foaled and kept at the Royal stable, and all of the animals were given personal names. At first, Frederick preferred to coin their names after their physical appearance or character, like Tiger, Fuchs, Brillant, Cerberus, or Sternrapp, but later he tended to name them after prominent politicians, military officers, high ranking generals, or great leaders both of antiquity and modern times.[45] His horses were now called Caesar, Choiseul, Brühl, or Kaunitz. Two horses, Pitt and Bute, bore the names of important British prime ministers – William Pitt, 1st Earl of Chatham, and John Stuart, 3rd Earl of Bute – who played important roles as allies of Prussia during the Seven Years' War. Bute's premiership was notable for the negotiation of the 1763 Treaty of Paris, which concluded a war that had almost destroyed the state of Prussia.

Some of Frederick's favourite horses were purchased from the best stables of England. In fact, his last and best loved horse, Condé, was born in an English stud farm in 1766 and sold eleven years later to the Prussian monarch. Condé was a gelding, valued as such by the monarch because it was quiet, lacked mating urges, and was less likely to be at odds with other horses. When the king first rode this English horse in 1777, he instantly liked it and decided to buy it at once. He named it after one of his French military heroes, Prince Louis

44 Louis A. DiMarco: War Horse. A History of the Military Horse and Rider. Yardley/PA 2012, p. 150. – In a letter of 2 October 1745, two days after the Battle of Soor, Frederick first reported to his valet Fredersdorf the sad loss of his horses Annemarie and Champion – only then he mentioned the death of his officers: Die Briefe Friedrichs des Großen an seinen vormaligen Kammerdiener Fredersdorf. Ed. by Johannes Richter. Berlin 1926, p. 55.

45 Cf. Sibylle Prinzessin von Preußen/Friedrich Wilhelm Prinz von Preußen, Friedrich der Große, p. 77.

FIGURE 4.5 Frederick (without spurs) on his favourite horse Condé. Oil on canvas. [Anonymous artist] (18th c.)

Henri de Bourbon-Condé, who was one of the most courageous and pre-eminent generals of seventeenth-century France. In addition, Condé was an important patron of the arts and of French literature and he, like Frederick, admired Molière as one of the greatest writers in the French language and in world literature.

The horse Condé enjoyed as many liberties as did the king's favourite dogs. It could not only roam freely in the park of Sanssouci but was also allowed to walk into the castle at any time. When Condé once made an excursion indoors, its heavy weight destroyed a large number of precious floor tiles in a central salon.[46] The king paid for the damage and smiled. One of the most detailed descriptions of Condé's special relationship with Frederick was given in a report from the Marchese Girolamo Lucchesini. He had made his way to Berlin in 1779, where he was hired as the monarch's lector and personal secretary at

46 Klaus-Dieter Budras and Rolf Berg: Einmal quer durch den runden Salon von Sanssouci. Condé – das letzte Leibreitpferd Friedrich II [!] von Preußen als Zeitzeuge der Geschichte Preußens und der Veterinärmedizin in Berlin. In: Reiten und Zucht in Brandenburg 1 (1998), p. 20–22.

first sight – because one of the king's Italian greyhounds took an instant liking to that elegant Italian diplomat. Lucchesini was the amused witness of a scene where, apparently, the horse dragged tasty melons and sweet figs out of the king's pockets while the monarch was standing next to the animal and did not notice a thing.[47]

On 4 July 1786, only a few weeks before he died, Frederick rode his favourite horse for the last time. It was a conscious and moving farewell. Condé survived the Prussian king for almost two decades and spent his final years in the grounds of the Royal Veterinary School in Berlin. When the horse died in 1804, its complete skeleton was carefully preserved and put on show in the exhibition of the Royal Anatomical Museum. More than two hundred years later, the skeleton of Frederick's last personal equine ('Leibreitpferd') can still be visited in Berlin. It has now become a rather bizarre attraction of the Free University, where since 1995 it has been on display in a special showcase in the Anatomical Museum of the Institute for Veterinary Anatomy.[48]

To this day, the bleached remains of the king's favourite horse bear witness to Frederick's deep conviction that there is not much difference between the lives and deaths of humans and animals. The horse's skeleton can be seen as a *vanitas* symbol of the fleeting and fragile existence which is our common fate as mortal beings. This symbolism was precisely the reason why Frederick – the *roi-philosophe* and poet of animals – wanted to be buried next to the bones of his dogs, in a fashion that was as eccentric as it was wise and humble.

Frederick the Great loved his animals and they were indeed dear to him. He was fond of dogs, horses, parrots, and monkeys, because they did not know how to lie and to flatter. They behaved freely and naturally, and when they wanted to be close to the king, they followed their true and unfeigned instincts. Animals took him, the powerful Prussian majesty, for what he was, an often very lonely man, who regularly sought to isolate himself from the world in his library. The king was extremely sceptical of the merits of humans – and he was self-critical enough to not exempt himself from the large number of men who had committed grave errors in their lives. A life shared with animals was for him an experience that was both humbling and comforting, a way to prevent himself from overestimating his intellectual strength and from falling into hubris.

47 Joachim von Kürenberg: Der letzte Vertraute Friedrichs des Großen. Marchese Lucchesini. Berlin 1933, p. 35.

48 Cf. Budras and Berg, Einmal quer durch den runden Salon, p. 20–22.

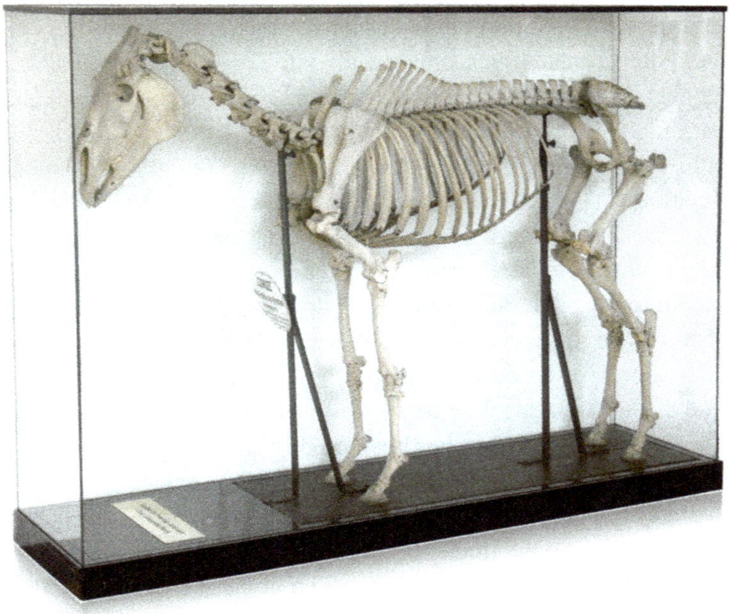

FIGURE 4.6 Taxidermied skeleton of Condé (1766–1804), the last horse of king Frederick the Great

CHAPTER 5

The Invention of the '*Cheval-machine*' as a Medical Response to the Machine Paradigm of the Enlightenment: Samuel Theodor Quellmaltz in Context

Stefanie Stockhorst

Résumé

En 1735, le professeur de médecine de Leipzig Samuel Theodor Quellmaltz (1696–1758) a conçu et fabriqué un cheval artificiel. Il l'a présenté dans un manuel de construction illustré avec des informations précises sur les matériaux et les dimensions de ce cheval en bois à usage thérapeutique. Cette contribution analyse l'invention du 'cheval-machine' par Quellmaltz en tant que contribution médicale et technologique au paradigme des machines au siècle des Lumières.

Abstract

In 1735, the Leipzig professor of medicine Samuel Theodor Quellmaltz (1696–1758) designed and built an artificial horse. He presented it in an illustrated construction manual, which included precise information about the materials and dimensions of this wooden horse for therapeutic use. This contribution analyses Quellmaltz's invention of the 'machine horse' as a medical and technological contribution to prevalent theories about the paradigmatic role of the machine in Enlightenment thought.

1 Humans and Animals in the Machine Discourse of the Enlightenment

Interest in machines was found before the eighteenth century; but in this period it intensified into enthusiasm. People appreciated not only useful machines that could take over certain functions, for example in agriculture or construction, but also the sophisticated apparatuses that could be admired in the art cabinets of early modern European courts. These machines impressed

 | DOI:10.1163/9789004495395_006

onlookers not necessarily by their practicality, but rather by their appearance, which was as artistic as it was complicated, with many surprising mechanisms. In the 'anthropological' eighteenth century, moreover, the fascination with machines encouraged a new approach to human self-interpretation. The Enlightenment machine paradigm offered one of many possibilities to describe the nature and functions of the *animal rationale*.

Of particular influence were the thoughts of René Descartes (1596–1650), according to whom "le Corps n'est autre chose qu'vne statuë ou machine de terre".[1] He insisted on negotiating the human body in a purely mechanical way: "Ie desire, dis-ie, que vous consideriez que ces fonctions suiuent toutes naturellement, en cette Machine, de la seule disposition de ses organes, ne plus ne moins que font les mouemens d'vne horloge, ou autre automate [...]."[2] Animals were also covered by this mechanistic understanding, but with certain constraints. In a letter to the English courtier and renowned horse-breeder William Cavendish, 1st Duke of Newcastle (1592–1676), for example, Descartes declared that he considered the behaviour of animals, as deliberate as it sometimes seemed, to be purely mechanical ("ainsi qu'vne horloge") and without any capacity for thought ("sans aucune pensée").[3] It seems that no explicit response from Cavendish has survived; but it is evident from his writings on horsemanship that he at least treated horses not like machines, but with gratitude and appreciation: "C'est la creature, entre toutes les autres, à qui l'homme a le plus d'obligation, tant pour l'vsage, que pour le plaisir, & tant pour son honneur, que pour sa vie; comment donc peut-on faire trop grande estime d'vn bon cheval?"[4]

The mechanistic body image was particularly influential in the field of (human) medicine: While the Cartesian model of the *animal-machine* already had philosophical opponents such as Pierre Gassendi and later Étienne Bonnot de Condillac in the seventeenth century, mechanistic approaches to explain the human body, its diseases and therapies[5] dominated in medical circles

1 René Descartes: Traité de l'Homme. In: idem: Œuvres. Ed. by Charles Adam and Paul Tannery. Vol. XI. Paris 1974, p. 119–215, here p. 120.

2 Ibid., p. 202.

3 René Descartes: Descartes au [Marquis de Newcastle]. [Egmond, 23 novembre 1646]. In: idem: Œuvres. Ed. by Charles Adam and Paul Tannery. Vol. IV. Paris 1976, p. 568–577, here p. 574–575.

4 William Cavendish, 1st Duke of Newcastle: AV ROY DE LA GRANDE-BRETAGNE. In: idem: LA METHODE NOUVELLE. Invention extraordinaire de dresser les Chevaux les travailler selon la nature, et parfaire la nature par la subtilité de l'art. Antwerpen 1658, [12 p., unpag.].

5 Cf. Sergio Moravia: From *homme machine* to *homme sensible*. Changing Eighteenth-Century Models of Man's Image. In: Journal of the History of Ideas 39 (1978), no. 1, p. 45–60, esp. p. 46–49; see also Catarina Caetano da Rosa: Androiden als Spie(ge)l der Aufklärung. Hamburg 2020, p. 48–51.

especially with the surge of fresh interest in the human condition. That key development has been termed the 'anthropological turn', dating from about 1750 in the German-speaking countries.[6]

In the course of the eighteenth century, many mechanics took up the challenge of imitating individual human functions by means of automata devoted to writing, calculating, playing chess or making music.[7] Samuel Theodor Quellmaltz (1696–1758), for his part, was committed to the construction and use of an artificial, but not automated, horse. The Leipzig professor of anatomy and surgery was one of the technically proficient advocates of the Enlightenment machine paradigm in Germany. In 1735 he published a short treatise entitled *Anweisung zur einer der Gesundheit dienlichen neu erfundenen Art der Bewegung*, in which he described a 'riding machine' for therapeutic purposes, taking into account the high costs involved in riding real horses.

Quellmaltz's invention originated in early modern mechanistic medicine, refined by the hydraulic notions with which William Harvey had explained the systemic circulation of blood in his ground-breaking study *De Motu Cordis* (1628).[8] At the same time, this '*cheval-machine*', which Quellmaltz himself referred to as his 'Pegasus',[9] marked the beginning of a trend towards an increased use of exercise equipment in physical training and medical therapy,

6 Cf. Carsten Zelle: Sinnlichkeit und Therapie. Zur Gleichursprünglichkeit von Ästhetik und Anthropologie um 1750. In: idem. (ed.): Vernünftige Ärzte. Hallesche Psychomediziner und die Anfänge der Anthropologie in der deutschsprachigen Frühaufklärung. Tübingen 2001 (Hallesche Beiträge zur europäischen Aufklärung Vol. 19), p. 5–24.

7 Cf. Adelheid Voskuhl: Androids in the Enlightenment. Mechanics, Artisans and Cultures of the Self. Chicago and London 2013; as well as Simon Schaffer: Enlightened Automata. In: idem, William Clark and Jan Golinski (ed.): The Sciences in Enlightened Europe. Chicago 1999, p. 126–165.

8 Katrin Löffler: Der Hometrainer. Eine Leipziger Erfindung? Vor 250 Jahren starb Samuel Theodor Quellmaltz, Konstrukteur einer 'Reitmaschine'. In: Leipziger Blätter 52 (2008), p. 62–63, here p. 62–63.

9 Cf. [Samuel Theodor Quellmaltz]: D. SAM. THEOD. QUELLMALZII. Anat. & Chir. P. P. E. Lips. NOVUM SANITATIS PRÆSIDIUM EX EQUITATIONE, MACHIÆ BENEFICIO INSTITUENDA. oder [!] Anweisung zur einer Der Gesundheit dienlichen neu erfundenen Art der Bewegung. Nebst dazu gehörigen Kupfern. Leipzig 1735, p. 12 (quotations from this text are referenced hereinafter with the abbreviation Q and corresponding page number); as well as Ehrendenkmahl des weiland *Magnifici*, Hochedelgebohrnen Hochachtbaren und Hocherfahrnen Herrn, HERRN *D.* Samuel Theodor Qvelmalz Erb-Lehn und Gerichts-Herrn auf Halß, der Arzneykunst ersten ordentlichen Lehrers, der medicinischen Facultät Dechant, der Academie Decemvir, des großen Fürsten-Collegii Collegiat etc. etc. welches Demselben durch eine auf der Universität zu Leipzig am 10ten Hornung 1759 öffentlich gehaltene Lobrede gestiftet hat HERR Carl Christian Krause der Weltweisheit und Arzneykunst Doctor, der medicin. Facult. Assessor. Aus dem Lateinischen Übersetzt von George August Friedrich Zeisig. Leipzig [1759], p. 29.

which was termed 'machine exercising'.[10] The objective of this contribution is to reconstruct the aims and effects of Quellmaltz's design of an artificial therapy horse, in order to locate his project as part of an enlightened discourse on the *machina animalis*.

2 Quellmaltz, Medicine, and Mechanics

Details of the life of the physician Quellmaltz are recorded in a Latin obituary, which was printed in German translation one year after his death. The appendix included a 66-title list of publications,[11] which covered a wide range of topics, most of them very short observations and dissertations on anatomy, physiology, surgery and pharmacology. Occasionally, rather special subjects such as a comment on a 'marine unicorn', which was brought to Leipzig as a zoological curiosity, were also to be found.[12] In addition, it was reported that Quellmaltz, in common with many scientists of his era, had interests in alchemy. He thus possessed extensive knowledge "of the secret journey and the hidden paths of this uncertain art".[13] In addition to the *Anweisung*, the fact that Quellmaltz provided the 7th edition of Gottfried Rothe's *Gründliche Anleitung zur Chymie* (1750; 1st ed. 1717) with comments and explanations is probably one of the main reasons for his popularity. It was his only other publication in German (however, it is so heavily interspersed with Latin that readers also needed a solid knowledge of that language). The *Anweisung* itself was bilingual, written in German and Latin, signifying that Quellmaltz not only wanted to reach the inner circle of his Latin-speaking colleagues, but also a wider readership.

Speculative 'knowledge' about the medical benefits of riding was already circulating in early modern conversational literature, which was supported by unconfirmed references to ancient authorities such as Aristotle and Pliny. For example, the polymath Erasmus Francisci (1627–1694), among others, supplied the following information:

10 Cf. Henning Eichberg: Stoppuhr, Reck und Halle. Zur Technisierung der Leibesübungen im 18. und frühen 19. Jahrhundert. In: Ulrich Troitzsch (ed.): Technologischer Wandel im 18. Jahrhundert. Wolfenbüttel 1981 (Wolfenbütteler Forschungen Vol. 14), p. 155–174, for Quellmaltz and other motion machines see p. 160–161.

11 Krause, Ehrendenkmahl, p. 35–36.

12 Samuel Theodor Quellmaltz: [De] vnicornu marinum ex vicinia Bremensi huc [sc. Lipsiam] delatur. In: Commercivm Litterarivm ad Rei Medicae et scientiae Natvralis 6 (1736), p. 171–173.

13 In the original: "auf der geheimen Reise und auf dem verborgenen Pfade dieser ungewissen Kunst" (Krause, Ehrendenkmahl, p. 25).

> Even though Aristotle writes that constant riding promotes lust, and even though some doctors advise those who are too cold and weak for sexual activity that they should ride often to warm themselves, all this is related to a moderate, gentle ride that is very good for both the stomach and the loins, as Pliny (lib. 18, c. 4) testifies.[14]

In the course of the seventeenth century, horseback riding increasingly found its way into the professional medical discourse. Influential authors such as Vopiscus Fortunatus Plempius (1601–1671), Georg Ernst Stahl (1659–1734), Herman Boerhaave (1668–1738) as well as his commentator Gerard van Swieten (1700–1772), Francis Fuller (1670–1706) and Johann Gottlob Krüger (1715–1759) all included the health-promoting effects of riding in their dietetic considerations on exercise. Movement itself was deemed beneficial, and it was also thought to be a promising cure, especially for digestive and circulatory problems, but also for various other diseases.[15] Occasionally, the healing effects of riding were discussed separately in the framework of academic disputations. Before the publication of Quellmaltz's *Anweisung*, this general point was made by Stahl in Halle with his *De novo specifico antiphthisico, equitatione* (1699), as well as by Michael Penz in Altdorf with his *Equitationis utilitates et incommoda* (1708) and by Didericus van Heemskerk in Leiden, with his *De equitatione* (1734).

Particular mention should be made of the academic disputation between Christian Michael Adolphi and Carl Friedrich Breitenbach, which was held in Leipzig on 17 March 1713 and then reported in print. Quellmaltz was 16 years old at that time and was to begin his studies three years later. However, this disputation appeared several times in Leipzig as *Dissertatio academica de equitationis eximio usu medico*. There were at least two versions in 1713 (one of them with a German translation of the title: *Von dem sonderlichen Nutzen des Reitens in vielen Kranckheiten*), and the text was then re-issued again in 1729 and 1744.

14 In the original: "Denn ob gleich Aristoteles schreibt/ das stetige reiten diene zur Geylheit: auch einige Aerzte denen/ die sich zu ehelichen Werck gar kalt und schwach befinden/ diesen Rath geben/ daß sie offt reiten/ und sich dadurch erwärmen sollen: ist doch solches alles/ von einem mässigem [!]/ sanfften/ und gelindem Ritt/ zu verstehen: welche beydes dem Magen und den Hüfften sehr gütlich thut: wie Plinius (lib. 18, c. 4) bezeuget." (Francisci, Erasmus: Das Alleredelste PFERD Der Gantzen Welt: Vermittelst eines anmutigen und erbaulichen Gesprächs/ Welches ist/ dieser Art/ die Eilffte/ Und zwar eine Winter=Monats=Unterredung. Frankfurt 1670, p. 277). – All translations given in this contribution are my own.

15 A synopsis of numerous positions is given by Herbert Hänel: Deutsche Ärzte des 18. Jahrhunderts über Leibesübungen. Frankfurt/M. 1972 (Studientexte zur Leibeserziehung Vol. 9), p. 119–123.

Such a long-lasting reception was a rare exception in the genre of the *dissertatio*, so it seems likely that Quellmaltz knew this survey, even though he did not mention it specifically.

As far as the medical value of riding via its particular mechanics of movement was concerned, Quellmaltz referred in the first place to the Italian physician Giorgio Baglivi (1668–1707), whose *Zwei Bücher De* PRAXI MEDICA had been available in German translation since 1718. Baglivi wrote that "the human body, according to its natural functions, is in reality nothing other than a confluence of chemico-mechanical movements which nevertheless derive from purely mechanical principles".[16] He advocated a 'newer' theory, based on "the laws of mechanics, the physical bodies and movement". It was therefore "able to resolve the incidences of disease far more successfully and safely than the theory of the Galenians [sc. followers of Galenic medicine]".[17] In a medicine that was committed to the *principium motus*, the bodily mechanics had to remain flexible, and blockages as well as stagnations had to be resolved.

With this concept in mind, Quellmaltz formulated the observation that was central to his recommendation for therapeutic riding: "Some of the juices in our bodies also have the property that they dry out or clot very easily if they are not kept constantly in motion by the external pressure in the above-mentioned manner".[18] For him, the traditional humoral pathology still offered orientation for the understanding of body fluids and temperaments. Therefore he referred, for example, to the recommendation of playing ball games, made by Galen of Pergamon (2nd c.), the highly influential physician who served in ancient Rome (cf. Q, p. 14; see also Galen: *De parvae pilae exercitio*) as well as to dietary instructions by the Byzantine physician Paulus of Aegina (7th c.) (cf. Q, p. 46).

Yet Quellmaltz took a much broader view in his treatise, referring to contemporary considerations on the flowing mechanics of body fluids (cf. esp. Q, p. 18–27), from which he derived numerous possible applications for movement

---

16 In the original: "der menschliche Cörper nach seinen natürlichen Verrichtungen in Wahrheit nichts anders sey/ als ein Zusammenfluss von *chymico-mechani*schen Bewegungen/ so dennoch von *pur mechani*schen *Principiis* herrühren" (Georgius Baglivius: Zwey Bücher De PRAXI MEDICA, Wie solche in klugen Observiren, wiederum auf den alten Fuß unserer fleißigen Vorfahren möge gestellet werden. Leipzig 1718, p. 180).

17 In the original: "auf den Gesetzen der *Mechanices*, der *Figur* und Bewegung fusset", "die Zufälle der Kranckheiten weit glücklicher und gewisser auffzulösen weiß/ als die *Theoria* der *Galen*isten" (Baglivius, Zwey Bücher De PRAXI MEDICA, p. 201).

18 In the original: "Einige von denen Säfften unsers Leibes haben über dieses diejenige Art, daß sie gar leichte vertrocknen oder geliefern [sc. gerinnen], es sey dann daß sie durch den angegebenen äusserlichen Nachdruck beständig in ihrer Bewegung erhalten werden." (Q, p. 22).

therapy using his artificial horse (cf. Q, p. 28–39). Such therapy would have a healing effect on the internal organs, the diaphragm, the gall bladder, the processes of respiration and digestion, and all areas wherever "the circulation of moisture is difficult".[19] Due to the holistic effects of equitation upon the body, "headaches, inflammations of the eyes and ears, toothache, tiredness, etc."[20] could also be cured.

Baglivi, quoted by Quellmaltz, commented on riding by reference to Stahl,[21] who advocated this type of exercise for colic, consumption, uterine and splenetic complaints, and constipation. Baglivi himself also drew upon both contemporary and ancient authors to link mechanistic medicine with the holistic doctrine of a regulated lifestyle. At any rate, patients should begin their exercises moderately:

> One starts by riding only half a quarter of a mile on a field path, but then continues this exercise for many miles, mainly in the morning hours when the weather is nice and bright. As the famous man Thomas Sydenham highly testifies. Long before him also Hippocrates *lib. de Diæta*.[22]

In an academic manner, Quellmaltz also hinted at a discussion of different opinions on the subject, in order to display his scholarly knowledge. Like Baglivi, he agreed with Thomas Sydenham (1624–1689), who cited riding on horseback as a remedy for gout.[23] Moreover, he declined Boerhaave's view that riding could be replaced "by riding in a carriage or by strolling".[24] Conversely, however, Quellmaltz also rejected the position of Friedrich Hoffmann (1660–1742), who wrote all too optimistically: "Horse riding surpasses the powers of all drugs, which are usually effective against consumption, hectic, scurvy and

19 In the original: "irgends der Umtrieb derer Feuchtigkeiten beschwerlich ist" (Q, p. 35).

20 In the original: "Kopffweh, Augen= und Ohren=Entzündungen, Zahnweh, Neigung zum Schlaff etc." (Q, p. 39).

21 Cf. Baglivius, Zwey Bücher De PRAXI MEDICA, p. 199.

22 In the original: "Den Anfang macht man/ daß man nur ein Feldweges/ halbe viertel Meyle reitet/ hernach aber diese Ubung auf viel Meylen *continuir*et und solches vornehmlich in den Morgen=Stunden bey schönen und hellen Wetter verrichtet. Wie solches der weltberuffene Mann *Thomas Sydenham* höchlich bezeuget. Lange vor ihm aber auch *Hippocrates lib. de Diæta*." (Baglivius, Zwey Bücher De PRAXI MEDICA, p. 215).

23 Cf. Q, p. 40; see also Thomas Sydenham: Tractatus de Podagra et Hydrope. London 1683, p. 99–100.

24 In the original: "durchs fahren in einer Gutsche oder durchs herumspatzieren" (Q, p. 41). See Herman Boerhaave: APHORISMI DE COGNOSCENDIS ET CURANDIS MORBIS IN USUM DOCTRINÆ DOMESTICÆ DIGESTI. 3rd ed., Leiden 1727 [1st ed. 1709], p. 188–189. Quellmaltz incorrectly refers to § 1281 instead of § 858.

cachexia."[25] Although Quellmaltz thought highly of therapeutic riding, he only wanted to put it "alongside" the "drugs".[26] When he put therapeutic riding into perspective, he was even a bit polemical, urging it as extremely beneficial, but not a universal remedy. Thus he scolded the lack of integrity of those colleagues who, "when they praise the usefulness of riding, call forth, I do not know what vapours that should come out of a magnanimous horse and which the rider should absorb again".[27]

And the technical side? For the construction of his riding machine with its different variants, Quellmaltz might have received inspiration from the enlightened mechanic Jakob Leupold (1674–1727), who spoke in favour of a greater orientation towards the practical application of academic mechanics. His workshop was locally based in Leipzig; and he gained a great reputation with his technical encyclopaedia *Theatrum Machinarum*. The first seven volumes (1724–27) were not only published while Leupold was still alive but were also chronologically significant for Quellmaltz's preliminary considerations about his own riding machine, which lasted for "about 9 years".[28] In the first volume, Leupold wrote about humans and animals which are conceived of as 'machines':

> The power of life or of humans and animals arises when the body, which is composed of muscles, with its blood and arteries, nerves, fibres and bones like a machine, is controlled by the soul in the same way as a machine is controlled by an external force, so that it can in turn move another body or machine through the hands, feet, legs and other parts of the body, and this is also after the machine of the body has been procured or is used with benefit.[29]

25 In the original: "Das Reuten übertrifft aller derer Medicamenten Kräffte welche sonsten bei der Schwindsucht, *Hectica*, *Scorbut*, und *Cachexie* alles ausmachen." (Q, p. 42). – See Friedrich Hoffmann: MEDICINÆ RATIONALIS SYSTEMATICÆ TOMVS PRIMVS. Halle 1718, p. 401 (no. XV).

26 In the original: "Artzneyen", "an die Seite" (Q, p. 42–43).

27 In the original: "wenn sie den Nutzen des Reutens loben, ich weiß nicht was vor Ausdünstungen, welche von einem großmütigen Pferde heraus dringen sollen, und welche der Reuter wieder in sich zu ziehen hätte, angeben" (Q, p. 15).

28 In the original: "ungefehr 9 Jahr" (Q, p. 7).

29 In the original: "Die Krafft des Lebens oder der Menschen und Thiere entstehet, wenn der Cörper, der aus Musculn, mit seinem Blut und Puls-Adern, Nerven, Fasen [!] und Beinen, als eine *Machine* zusammen gesetzet ist, durch die Seele, als wie eine *Machine* durch die äusserliche Krafft regieret wird, daß sie durch die Hände, Füsse, Beine und andere Theile des Leibes einem [!] andern Cörper oder *Machine* wieder bewegen kan, und zwar auch nachdem die *Machine* deß Cörpers beschaffen, oder mit Vortheil gebrauchet wird." (Jacob

As he saw natural limits in the horse's capacity and willingness to perform, in contrast with a tireless machine, Leupold thought not only about the sensible use of a machine, but also about its technical substitutability[30] – reflections that were also highly relevant for Quellmaltz.

3 The Riding Machine: Advantages, Variants, and Usage

Quellmaltz evidently saw himself not only as a professor of medicine, but also as a practising physician, which shaped his approach in the *Anweisung*. Essentially, he abstained from philosophical reasoning. Instead, he argued as a practitioner who observed that physical exercise keeps a person in good health. Although he did not base his statements on specific case histories, he consistently referred to the course of diseases, as well as suitable therapies, based upon evidence from his own wealth of experience as a doctor. On several occasions, he had met "such sick people" who expected "help from riding".[31] He considered these patients in the context of general trends in the lifestyle of rural and urban populations. He contrasted the health status of these two groups, concluding that urban dwellers developed many ailments due to lack of exercise. Furthermore, Quellmaltz demonstrated not only a medical but also an ethical evaluation of physical effort. Thus he wrote about the "benefits of work" for promoting health, deploring the fact that many avoided it "out of leisureliness and wantonness, because of the accompanying discomfort".[32]

Among the various ways to gain physical health, Quellmaltz considered riding to be particularly curative. Quellmaltz was seemingly the first author to publish a 'popular' monograph as a special plea for therapeutic riding, in which riding was not just one out of many dietary recommendations. He justified the medically favourable effects of riding by the fact that it involved not only single parts of the body, but the entire body in the interaction of its parts. He noted an additional advantage in that the rider is set in motion in a three-dimensional way: up and down, forwards and backwards and swaying to both sides. Thus Quellmaltz explained:

Leupold: Theatrum Machinarvm Generale. Schau-Platz Des Grundes Mechanischer Wissenschafften. Vol. 1. Leipzig 1724, p. 114).

30 Cf. Leupold, Theatrum Machinarvm Generale, Vol. 1, p. 118–119, p. 159 and p. 239.

31 In the original: "solche Krancke wieder vorgekommen, welche sich von den [!] Reuten Hülffe" versprachen (Q, p. 7).

32 In the original: "Arbeit vor Nutzen", "aus Gemächlichkeit und Wollust, wegen der damit verknüpfften Beschwerlichkeit" (Q, p. 2).

> Among these [sc. physical exercises], riding, which consists in a movement that pushes the human body forward and backward, on a horse that moves the front and back legs either simultaneously or alternately, sometimes lifting, sometimes lowering them again, sometimes turning one side forward around the other, and sometimes alternately pushing the whole body forward and backward, has many advantages. […] The one described first, which is a type of shaking, is more effective than the others because of the greater movement of the body, and is also generally more beneficial to health than the others, although their benefits are not denied.[33]

Although riding was widely advocated as a form of therapy in eighteenth-century medicine, it was too expensive for many patients. The impetus for the development of a riding machine was therefore mainly based on the argument of cost:

> I wondered, then, whether this movement could not be imitated through art, all the more so because those who otherwise applied considerable diligence to the invention of new types of movement, this one which is in keeping with riding (if I exclude the ordinary plank horses of children, whose usefulness is certainly not to be derided) do not tell of it in a word?[34]

In addition to the cost savings in comparison with riding real horses, Quellmaltz cited several other advantages of his riding machine: It made riding independent of the prevailing weather conditions. At the same time, it was available

33 In the original: "Unter solchen [sc. physical exercises] nun hat das Reuten, welches in einer Bewegung bestehet, die den menschlichen Leib, auf einen [!] Pferde, das die forder und hinter Füsse entweder zugleich, oder wechselweise fortsetzet, bald in die Höhe hebt, bald wieder niederläst, bald eine Seite um die andere vorwerts wendet, bald aber auch wechselweise den gantzen Cörper vor und hinterwerts stöst [!], vieles zu [!] voraus. […] Die zuerst beschriebene, welche eine Art einer Schüttelung vorstellt, übertrifft wegen stärckerer Bewegung des Cörpers die übrigen, sie ist auch der Gesundheit insgemein zuträglicher als die andern, denen aber ihr Nutzen deswegen keinesweges abgesprochen wird." (Q, p. 4).

34 In the original: "Jch dacht also bey mir selbst: ob nicht solche Bewegung durch die Kunst nachzuahmen stünde, und solches um desto mehr, je weniger diejenigen, so sonst in Erfindung neuer Arten derer Bewegungen nicht vergebenen Fleiß angewendet, dieser, so mit den [!] Reuten übereinkömmt (wenn ich der Kinder aus Bretern [!] gemachte gewöhnlichen Pferde, davon der Nutzen sicherlich nicht zu verachten ist, ausnehme,) nicht mit einem Worte Meldung thun können?" (Q, p. 6–7).

as required, even several times a day. The lack of strength and resilience of sick people could be taken into account. The artificial horse offered the further possibility of making this form of therapy suitable for women by installing a side-saddle, instead of the conventional saddle that was used astride. Indeed, Quellmaltz provided a separate illustration to demonstrate a side-saddle in position (figure 5.2). Riders of the artificial horse were not at the mercy of the "boldness" and "stiffness" of the "ordinary loan horses", and there was no danger of falling.[35] Such considerations provided the motivation to surpass the unpredictability of nature by developing a technical invention.[36] Moreover, riders did not have to "dress properly" to use it.[37] Last but not least, the machine was also suitable for children, because it could be adjusted to run very smoothly (cf. Q, p. 44).

A major focus in Quellmaltz's *Anweisung* lay in its precise, illustrated instructions for assembling the riding machine. It contained details of the materials and dimensions required for various construction options, in which a fur-covered wooden horse (figure 5.1) could be suspended from ropes, pulleys and sprung beams. If the available space was not sufficient for a whole horse, the alternative allowed a saddle (figure 5.3) to be mounted in the bracket instead. The accompanying illustrations show that space and wealth were correlated, for in the room containing the full version of the riding machine with a whole horse, a much more elaborate decoration of the walls also catches the eye. Basically two different ways of building the suspension were also envisaged. If the room chosen for the construction had wooden ceiling beams, the construction was to be attached to them, whereas in the case of plaster or stone ceilings, wall mounting or suspension in a free-standing wooden frame was suggested (cf. Q, p. 8) (figure 5.3). In all cases, it was recommended that the artificial horse should be mounted near the bedroom so that it would be easily accessible from there (cf. Q, p. 7–8).

In terms of using the machine, Quellmaltz stated that it could be set in motion both by the rider's body-weight or by a hand-operated cable pull. In detail, he explained:

> It is now easy to notice that the machine starts moving as soon as he [sc. the rider] sits on it. For when the bars are bent by the weight of the body and by the pressure exerted by its muscles, they lift the whole body up

35 In the original: "Muthwillen", "Steiffigkeit", "gewöhnlichen Mieth=Pferde" (ibid., p. 17).

36 Cf. Löffler, Der Hometrainer, p. 62.

37 In the original: "ordentlich ankleiden" (Q, p. 17–18). – The entire catalogue of benefits is also presented in [Anon.], [Art.] 'Motions-Maschine (Quelmaltzische [!])', col. 1940.

> again as soon as the pressure is released, or when the wheels and levers are released and come to rest by grasping and pulling the belt or cord. As soon as the strap or cord is released, however, the body overcomes them again and pulls them taut.[38]

As an additional advantage, the pulley could also be "handed over to a friend or servant" if the patient was too weak to activate the mechanism.[39]

As for the duration of rides on the riding machine, Quellmaltz gave guidelines deriving from humoral pathology. According to this approach, people who are "inclined to a lot of mucus" as well as those who "eat many and more nutritious foods" should ride correspondingly "longer and stronger".[40] He recommended using the machine either before meals or with a longer break for thorough digestion after a meal, "so that the crudities and impurities that have crept into the blood are expelled or improved".[41] The treatment could be undertaken "all at once for half an hour without fear of exhaustion", only "shorter perhaps in the case of particular physical weakness or at the beginning when they [sc. the patients] are not yet used to the machine".[42]

4 Building of Riding Machines

Probably the greatest literary fame among all artificial horses was achieved by the wooden horse which, according to mythological tradition, the Greeks used to gain entry to the city of Troy. There is also evidence of actual constructions of artificial horses in ancient times. In Roman military training, for

38 In the original: "Hierbey ist nun leicht zu erachten, daß so bald er [sc. der Reiter] sich darauf setzet, die Maschine sogleich in Bewegung komme. Denn wenn die Stangen durch die Schwere des Leibes und durch die von dessen behörigen Musculn erweckte geringste Druckung gebeuget werden, heben sie den gantzen Leib wieder in die Höhe, so bald sie bey Nachlassung benannter Druckung, oder auch bey Ergreif- und Anziehung des Riems [!], oder Schnure derer Räder und Hebel wiederum frey werden und in die Ruhe kommen. Hinwiederum so bald der Riem oder die Schnur nachgelassen wird, werden sie wiederum vom Leib überwunden u. gespannet." (Q, p. 12–13).

39 In the original: "einem Freund oder Diener übergeben warden" (ibid., p. 14).

40 In the original: "zu vielem Schleim geneigt", "viele und stärcker nährende Speisen zu sich nehmen", "länger und stärcker" (ibid., p. 45).

41 In the original: "damit die *Crudi*täten und Unreinigkeiten, so sich irgends in das Blut eingeschlichen, dadurch ausgeworffen oder verbessert warden" (ibid., p. 46).

42 In the original: "auf ein mahl ein halbes Stündgen ohne alle Besorgung einer Entkräfftung", "weniger vielleicht wegen besonderer Leibes-Schwachheit oder anfänglich, ehe sie [sc. die Patienten] noch recht der Maschine gewohnt" (ibid., p. 47).

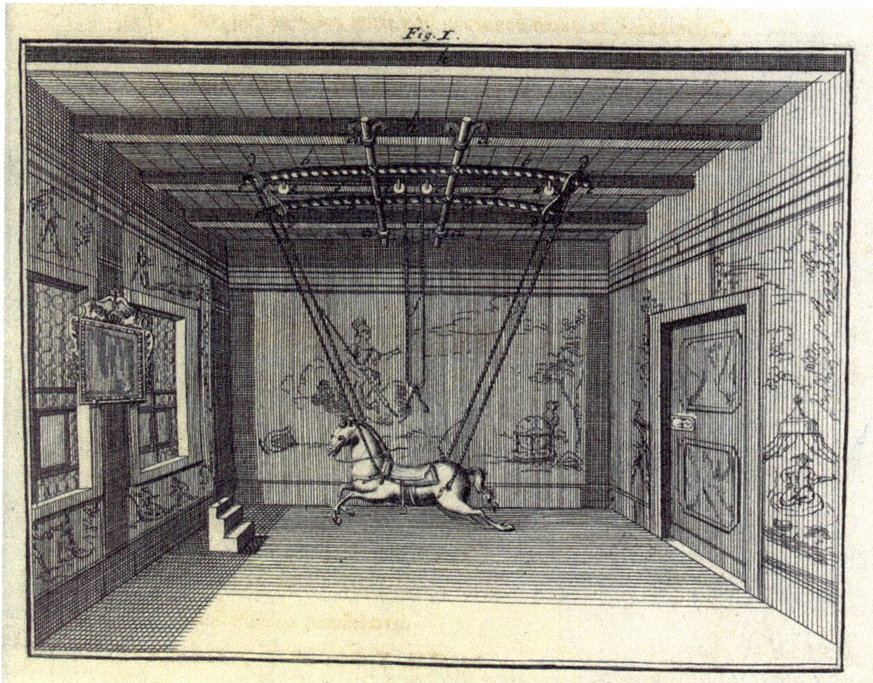

FIGURE 5.1 The 'elaborate' version of Quellmaltz's riding machine with a full, detailed horse with mounting on the ceiling

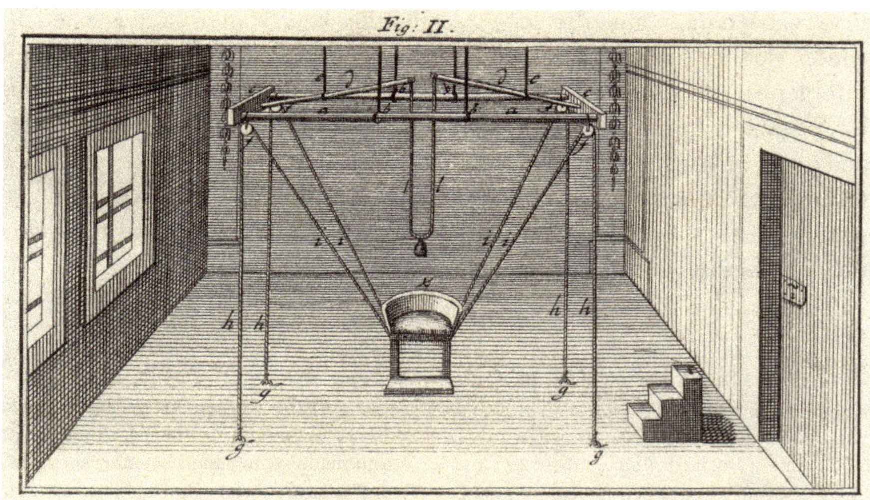

FIGURE 5.2 The 'ladies' version of Quellmaltz's riding machine with a side-saddle

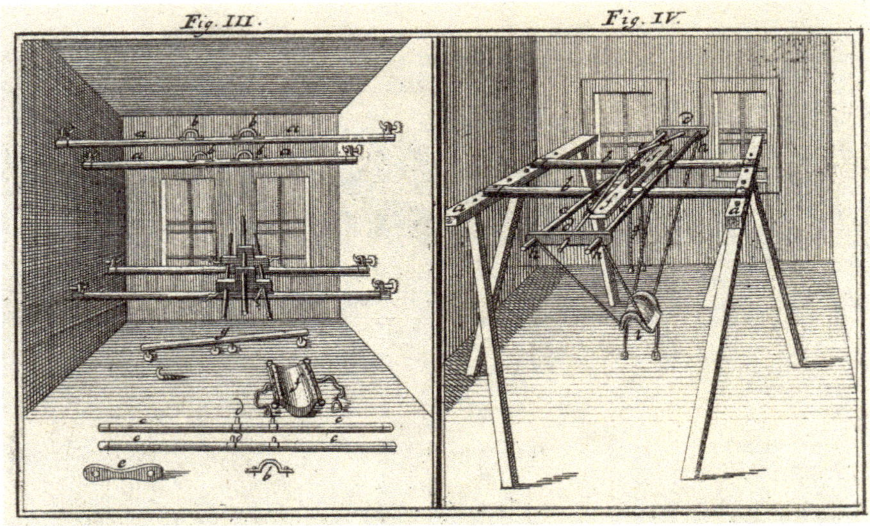

FIGURE 5.3 The 'low budget' version of Quellmaltz's riding machine with just a rocking saddle, left with mounting on the wall, right in a wooden frame

example, wooden horses were already used to train recruits on horseback (cf. Vegetius, *De re militari*, 4th c.).[43] Vegetius's report was translated into German in 1616 by Johann Jacob von Wallhausen, who in the same year advocated, in his *Ritterkunst,* the use of a life-size model horse standing firmly on wooden planks for early modern knightly training. His words ran as follows: "This practice has been maintained so far, although it has been somewhat neglected. In winter, wooden horses were kept under a roof (to protect them from snow and turbulence), in summer, they were placed in the open air: the first Tyrones [sc. beginners in the knightly profession] who began to ride had to jump up and down on them, first without weapons and then armed."[44] In several illustrations (figure iv, no. 1–3) Wallhausen demonstrated how this wooden horse was meant to be used to practise mounting and dismounting, whilst carrying unwieldy equipment.

43 Cf. Eichberg, Stoppuhr, Reck und Halle, p. 160.

44 In the original: "Welcher Gebrauch bißhero ist erhalten worden/ ob schon er etwas nachlässig ins Werck gesetzt würde. Deß Winters wurden hültzerne [!] Pferde vnder einem Dach (daß sie vor Schnee vnnd Vngestümme etwas möchten bewahrt seyn) deß Sommers vnder den freyen Himmel gesetzt: auff diesen müsten die ersten anfangende *Tyrones*, erstlichen ohn Waffen/ demnach mit Waffen gewapnet/ auff- und abspringen." (Johann Jacob von Wallhausen: Ritterkunst. Frankfurt/M. 1616, p. 73).

FIGURE 5.4 Wallhausen – knightly exercises with a stationary wooden horse (1616)

In his *Vollkommener ergäntzter Pferdt=Schatz* (1664), the German compiler Johann Christoph Pinter also recommended the use of an artificial training horse, which however did not stand immobile on its legs like Wallhausen's, but could swing on large skids.

This horse was described by Pinter as a 'Pluvinell-Pferd'[45] in homage to the French riding master Antoine de Pluvinel (1555–1620). Yet its construction was not described in the text, but only illustrated. Pinter was equally not advocating this exercise for therapeutic purposes. Rather, the basic riding training was to be supported by this wooden horse, in order to spare horses and riders until beginners had developed an elementary feeling for movement. On the one hand, the horses would be spared if the riding students were prepared accordingly. On the other hand, Pinter sought to train beginners by means of dry exercises on a kind of rocking horse for adults. In this way they could learn the necessary movements "without pain, annoyance, exertion, danger and damage in a short time".[46] Because a real horse would notice quickly when it was teamed up with a beginner:

> Which a horse can easily recognise from the timid appearance with the awkward and insecure mounting alongside the untidy wrong movements and knows how to take advantage of this and tries hard to get rid of the rider with constant use of his means of defence and his behaviour, and also reaches this goal all the sooner and more because the confusion of the rider is thereby doubled.[47]

As far as the functioning of these artificial horses was concerned, the main difference between the designs of Pinter and Quellmaltz consisted in the fact that the former had a static structure, while the latter consisted of reciprocally movable parts, with a mechanical drive system. Furthermore, the two artificial horses were intended for different purposes. The Leipzig medical doctor was

45 Cf. Pinter, Vollkommen ergäntzter Pferd=Schatz, Vol. 2, p. 10.

46 In the original: "ohne Schmertzen/ Verdruß/ Beschwerung/ Gefahr und Schaden/ in kurzer Zeit" (Johann Christoph Pinter von der Aue: Vollkommener ergäntzter Pferdt=Schatz. Vol. 2: Appendix deß andern Theils Deß ersten Haubt-Theils dieses Pferdt-Schatzes. 2nd ed., Vienna 1695 [1st ed. Frankfurt/M. 1664], p. 10).

47 In the original: "Welches so dann ein Pferd/ auß dem zaghafften Zusprechen/ sambt dem unrechten und unversicherten Auffsitzen/ neben den unordentlichen falschen Bewegungen/ leichtlich mercket/ und sich solches wol zu nutz zu machen weiß/ und mit beständigem Verfolg seiner Defensions=Mittel/ und Bezeigung sich des Reuters zu befreyen nicht unversucht lässet/ und auch um so viel ehe und mehr darzu gelanget: als auch hierdurch deß Reuters Verwirrung verdoppelt wird." (Pinter, Vollkommen ergäntzter Pferd=Schatz, Vol. 2, p. 13). – Pluvinel himself expresses similar concerns in his riding manual, but does not recommend wooden horses, but very well trained school horses, so that "le cheval bien dressé luy [sc. dem Reitanfänger] aprenne [!] tout qu'il est besoin." (Antoine de Pluvinel: L'instruction dy Roy en l'exercice de monter a [!] cheval. Paris 1625, p. 16).

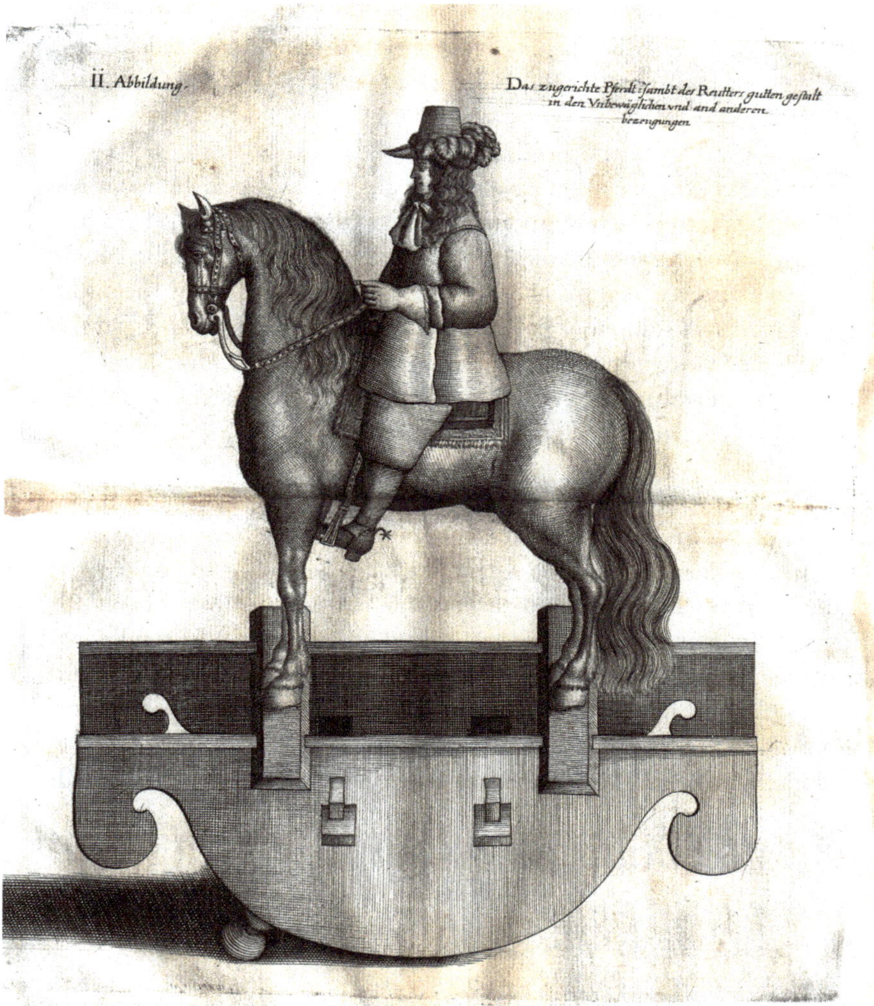

FIGURE 5.5 Pinter – wooden training horse for beginners in equitation (1664)

not interested in preliminary exercises for trainee riders, but in the machine's therapeutic use for various illnesses, caused by lack of exercise.

There are only very few surviving clues as to the actual distribution and use of Quellmaltz's riding machine. First of all, the unusually high number of surviving copies of his book suggests that it had some success – apparently the *Anweisung* has been preserved in several collections through the centuries. At present, 28 copies can be tracked worldwide, 16 of them in German and 12 in international libraries.[48] The obituary notice of Quellmaltz also reported on the superregional impact of the invention, reaching as far as Italy. Hence it does seem that it was actually built and used, even if there is no evidence to date of any surviving machines:

> This excellent invention has not only been of great benefit in our country and has earned its inventor many thanks, but has also made both him and our university famous even in the farthest reaches of Italy. It has been found to have proved its worth there through many experiments on people suffering from consumption, and has been considered worthy of being translated almost from word to word into the language of this country, and of being made known to doctors and patients throughout Italy as a safe remedy against consumption and other physical ailments.[49]

Based on this information, a review of this obituary was available to be read shortly afterwards: "His riding machine, invented for the best of health, has become popular not only among us but also in Italy."[50] As the only one of his brainchildren, Quellmaltz's riding machine also got its own entry in Johann Heinrich Zedler's *Universal-Lexicon*. Indeed, the report ran for a full four-and-a-half columns. It contained the essential technical details of the construction, the complete catalogue of its benefits and all the illustrations from the book.[51] Although no information was given about the diffusion of the artificial horse, Quellmaltz was explicitly credited as a scholar serving the common good:

---

48 With warm thanks to Luise Seifert for her support in this enquiry.

49 In the original: "Diese vortrefliche Erfindung hat nicht allein in unsern [!] Lande grossen Nutzen geschaft [!], und seinem Erfinder vielen Dank erworben, sondern auch in den äusersten Grenzen Jtaliens ihn sowohl als unsere Universität berühmt gemacht. Man hat es nehmlich daselbst durch viele Versuche an Leuten, die mit der Schwindsucht behaftet gewesen sind, bewährt gefunden, und es für würdig gehalten, es fast vom Wort zu Wort in die Sprache dieses Landes zu übersetzen, und als ein sicheres mittel wider die Schwindsucht und andere Leibesgebrechen den Aerzten und Kranken durch ganz Jtalien bekannt zu machen." (Krause, Ehrendenkmahl), p. 29.

50 In the original: "Seine zum Besten der Gesundheit erfundene Reitmaschine, ist nicht nur unter uns, sondern auch in Italien beliebt geworden." ([Anon.]: Memoria Sam. Theod. Quellmalz. In: Freymüthige Nachrichten van neuen Büchern und andern zur Gelehrtheit gehörigen Sachen 16 (1759), p. 393–394, here p. 394).

51 Cf. [Anon.]: [Art.] 'Motions-Maschine (Quelmaltzische [!])'. In: Grosses vollständiges Universal-Lexicon Aller Wissenschafften und Künste [...]. Ed. by Johann Heinrich Zedler. Vol. 21: Mi–Mt. Leipzig and Halle 1739, col. 1939–1943.

> The more necessary [...] physical activity is for the maintenance of health, the more praise is due to those who make an effort to show people, by means of useful inventions, the way in which they can either maintain their health through appropriate and comfortable motion, or, if they have lost it due to lack of motion, regain it.[52]

A later mention of the riding machine was made by Gerhard Ulrich Anton Vieth (1763–1836), a physical education specialist from Hooksiel, who, while working in Dessau, published *Versuch einer Encyklopädie der Leibesübungen* in 1795. Here he casually stated – interestingly not in the elaborate chapters on vaulting and riding, but in the section on swinging – that: "Among the pendulum swings, Quellmalz's riding machine can be counted as well, as it were."[53] The following year, there was a strikingly similar formulation in Krünitz's *Encyklopädie*, which can be explained either by a very rapid reception of the earlier text by the later one, or by transmission via an (unknown) common source text: "Among the pendulum-swings, Quellmalz's riding machine can be counted as well, as it were."[54]

In addition to the outlined tasks in riding training and therapy, artificial horses were also used in other fields of practice in the (long) eighteenth century, which for the sake of comprehensiveness should at least be mentioned here. Firstly, the 'bidet' developed for the washing of intimate parts was a construction inspired by horses: It was first mentioned by René Louis de Voyer, marquis d'Argenson (1694–1757), foreign minister under the French King Louis XIV. In his memoirs, d'Argenson accounts of how, as early as 1726, he found the influential mistress Jeanne-Agnès Berthelot de Pléneuf, marquise de Prie (1698–1727) in her private chamber while using a bidet.[55] At that time, these

52 In the original: "Je nöthiger [...] die Bewegung dem Menschen zu Erhaltung der Gesundheit ist, desto grösseren Lobes sind diejenigen werth, welche sich bemühen, den Menschen durch nützliche Erfindungen den Weg zu zeigen, wie sie durch eine hinlängliche und bequeme Motion ihre Gesundheit entweder erhalten, oder wenn selbige durch Ermangelung dieser Bewegung verlohren gegangen, selbige wieder erlangen können." ([Anon.], [Art.] 'Motions-Maschine (Quelmaltzische)', here col. 1939).

53 In the original: "Zu den Pendelschaukeln kann gewissermaßen auch die Quellmalzische Reitmaschine gerechnet werden." (Gerhard Ulrich Anton Vieth: Versuch einer Encyklopädie der Leibesübungen. Vol. 2. Berlin 1795, p. 110).

54 Oekonomisch-technologische Encyklopädie oder allgemeines System der Staats- Stadt-Haus- und Land-Wirthschaft, und der Kunst-Geschichte in alphabetischer Ordnung; von D. Johann Georg Krünitz. Vol. 17. Berlin 1796, p. 582.

55 Cf. Luciano Spadanuda: Storia del bidet. Un grande contenitore ideologico. Rom 1998, esp. p. 5–8; based on this with a confessional argument Roberto Zapperi: Zu viel Moralismus macht den Körper schmutzig. In: FAZ No. 95, 24 April 2010, p. Z3.

were elongated wooden stools with a metal or porcelain wash-bowl embedded in them. To perform intimate ablutions the user sat astride it, as on a small horse. Indeed, the name of this intimate utensil was based on this equine analogy, since it initially referred to a small riding pony, originating in France, known as the 'bidet'.[56] A painting of a bidet (c. 1790), by the French artist Louis-Léopold Boilly (1761–1845),[57] showed especially clearly the utensil's horse-like shape. As a 'gallant' insinuation "à l'ère du libertinage triumphant",[58] the associative link between riding and intimate washings may well have been a matter for suggestive laughter in raffish circles. On the other hand, a quite different mechanical vehicle was also called a 'riding machine', namely the single-track balance bike presented by Karl von Drais in 1817, which was also known as a 'draisine' after its inventor.[59] It was first developed into a 'velociped' and later into a bicycle.

5 Conclusion

Radical materialism, which reached its peak from Descartes's early approaches to Julien Offray de La Mettrie with *L'homme machine* (1747), was primarily concerned with metaphysical issues. Yet it had immediate consequences for the theory of riding later in the eighteenth century. In line with the machine paradigm of the Enlightenment, the French riding master Louis-Charles Mercier Dupaty de Clam chose to use mechanistic metaphors in his ground-breaking treatise *La Science et l'Art de l'Equitation* (1776) to describe the coordination of the movements of human and horse. For example, he wrote:

---

56 Cf. [Anon.]: [Art.] 'BIDET PETIT'. In: Grosses vollständiges Universal-Lexicon Aller Wissenschafften und Künste […]. Ed. by Johann Heinrich Zedler. Supplementary Vol. 3: Barc-Bod. Leipzig 1753, col. 1160. – See Jacques Mulliez: Du bidet au poney, du travial paysan au loisir aristocratique. In: Daniel Roche and Daniel Reytier (ed.): À cheval! Écuyers, amazones & cavaliers du XIVe au XXIe siècle. Paris 2007, p. 49–69.

57 Reproduced i. a. in Spadanuda, Storia del bidet, p. 26; see also Louis-Léopold Boilly: *La toilette intime*, also known as *The Morning Wash* (c. 1790); location unknown (cf. Étienne Bréton and Pascal Zuber (ed.): Boilly. Le peintre de la société parisienne de Loius XVI à Louis-Philippe [catalogue raisonné]. 2 Vols. Paris 2019, Vol. 2, p. 911, No. 2415 PM).

58 Fanny Beaupré und Roger-Henri Guerrand: Le confident des dames. Le bidet du XVIIIe au XXe siècle: histoire d'une intimitè. Paris 1997, p. 16; see here also for more detailed etymological explanations.

59 Cf. Johann Heinrich Moritz [von] Poppe: Encyclopädie des gesammten Maschinenwesens, oder vollständiger Unterricht in der praktischen Mechanik und Maschinenlehre. Vol. 7 (Supplementary Vol. 2): A–Z. Leipzig 1818, p. 400.

> Comme la construction d'une machine quelcon que doit être déterminée sur l'usage que l'on veut en faire: de même la position de l'homme se réglera sur l'emploi de ses membres à cheval, en composant les forces différents membres de maniere qu'il s'ensuive, de la part de l'homme, une direction fixe, &, de la part du cheval, une vîtesse connue: car l'Equitation résulte d'une quantité de puissances composées ensemble.[60]

Additionally, Dupaty for his part referred to a collection of self-built machine models, with the help of which one of his colleagues illustrated the mechanics of the art of riding: "M. [Hébert] DE LA PLEIGNIERE, Ecuyer du Roi à Caen, connu par ses talents en tout genre, possede une superbe collection anatomique, avec laquelle il démontre son Art, à l'aide de machines qu'il a composées lui-même pour le même sujet."[61]

Was Quellmaltz aware of the great changes in the history of thought, within which his text appeared? It must be admitted immediately that his writings do not reveal any philosophical ambition worth mentioning. Yet the biographical entry on Quellmaltz in Zedler's *Universal-Lexicon* stressed that the Leipzig professor was not only a "famous doctor of pharmacology" but also a "master of philosophy and liberal arts".[62] Of course, this qualification did not say very much about the actual depth of his philosophical interests, but rather tended to refer to his general academic education: In the university landscape of this era, disciplinary specialisations were only gradually becoming differentiated from universal scholarship. However, the obituary of Quellmaltz also mentioned his particular commitment to philosophical matters. His "untiring study of philosophy" was recorded as well as the fact that he attended "the lectures of the philosophers, as of Mr. Rüdiger and the two Mr. Müller, with particular diligence, for long and uninterruptedly".[63]

60 Louis-Charles Mercier Dupaty de Clam: LA SCIENCE ET L'ART DE L'EQUITATION, démontrés d'après la nature; OU THÉORIE ET PRATIQUE DE L'ÉQUITATION, fondées sur l'anatomie, la méchanique, la géometrie, & la physique. Paris 1776, [livre premier] p. 2f.

61 Louis-Charles Mercier Dupaty de Clam: PRÉFACE. In: idem: LA SCIENCE ET L'ART DE L'EQUITATION, démontrés d'après la nature; OU THÉORIE ET PRATIQUE DE L'ÉQUITATION, fondées sur l'anatomie, la méchanique, la géometrie, & la physique. Paris 1776, p. 16–24, here p. 21.

62 In the original: "berühmter Doctor der Artzneykunst", "Magister der Philosophie und freyen Künste" ([Anon.]: [Art.] 'Quellmaltz (Samuel Theodor)'. In: Grosses vollständiges Universal-Lexicon Aller Wissenschafften und Künste [...]. Ed. by Johann Heinrich Zedler. Vol. 30: Q und R–Reh. Leipzig and Halle 1741, col. 188–189, here col. 188).

63 In the original: "unermüdetes Studieren der Philosophie", "die Vorlesungen der Philosophen, als des Herrn Rüdigers und der beiden Herren Müller, mit besonderm Fleiße, lange und ununterbrochen" (Krause, Ehrendenkmahl, p. 23). – This refers probably to Andreas Rüdiger (1673–1731) and August Friedrich Müller (1684–1761); it is not clear who the other of the "Herren Müller" is.

Thus it may at least be assumed that Quellmaltz shared an intellectual participation with the historical developments of Enlightenment thought, which went beyond the proven knowledge of mechanistic medicine. With the design of his artificial horse, he reacted to what seemed to be medically adequate and technically feasible in his era. Therefore it seems plausible to conclude that he participated productively in the maelstrom of new ideas about machines, without himself contributing to their philosophical underpinning. It was not until a little later that Ernst Platner, also a medical doctor in Leipzig, positioned himself as an expressly 'philosophical' physician. With his *Anthropologie für Aerzte und Weltweise* (1772), he shaped the anthropology of the late Enlightenment in the German-speaking cultural sphere.[64]

By contrast, Quellmaltz, a medical doctor in academic service, made a name for himself as an 'inventor' with the very clear practical orientation of his riding machine. The presentation of his *Anweisung*, which encouraged the reproduction of his machine, may be perceived as a distant academic reflection of the broader 'projecting age'. But beyond the fact that he praised the usefulness of a construction manual for an artificial horse, Quellmaltz did not meet the common definition of a 'projector'. This social type is often described as culturally marginal, busily absorbed in day-to-day affairs, and willing to cut corners. Quellmaltz did not fit that model, in that his inventive enterprise was not designed for a long production run and the pursuit of profit.[65]

Instead, his contemporaries regarded Quellmaltz as a representative of the Enlightenment ideals of utility and the common good, who was dedicated to using his powers for the betterment of the human condition.[66] Accordingly, the obituary paid tribute to the fact that his works were characterised by "useful remarks" and "advice" that was aimed at serving "the common good".[67] He was

64 Cf. Moravia, From *homme machine* to *homme sensible*, p. 49–51; see in detail Alexander Košenina: Ernst Platners Anthropologie und Philosophie. Der ‚philosophische Arzt' und seine Wirkung auf Johann Karl Wezel und Jean Paul. Würzburg 1989 (Epistemata; Series Literaturwissenschaft Vol. 35), esp. ch. 1.

65 Cf. for a definition of the 'project maker' as a phenomenon of the Enlightenment Georg Stanitzek: Der Projektmacher. Projektionen auf eine ‚unmögliche' moderne Kategorie. In: Ästhetik und Kommunikation 17 (1987), no. 65/66, p. 135–146 [again in: Markus Krajewski (ed.): Projektemacher. Zur Produktion von Wissen in der Vorform des Scheiterns. Berlin 2004 (Copyrights Vol. 15), p. 29–48]; as well as Jan Lazardzig: Projektemacher als Virtuosen des Wissens? In: Gabriele Brandstetter, Bettina Brandl-Risi and Kai van Eikels (ed.): Prekäre Exzellenz: Künste, Ökonomien und Politiken des Virtuosen. Freiburg [et al.] 2012 (Rombach Wissenschaften; Series Scenae Vol. 14), p. 37–55.

66 See also Löffler, Der Hometrainer, p. 63.

67 In the original: "nützliche Anmerkungen", "Rathschläge", "zum gemeinen Besten" (Krause, Ehrendenkmahl, p. 28–29).

not motivated by money. His personal lifestyle was ascetic,[68] while his prime medical commitment was a disinterested care for others: "He was beloved by everyone, kind to everyone, especially to the poor, whom he helped without payment, giving them the most salutary medicines free of charge."[69]

Quellmaltz thus ranks among the pioneers of therapeutic riding. This research suggests that he was the first author to publish a 'popular' monograph (written in the German vernacular language and thus widely readable), which conveyed a public plea for therapeutic riding. That exercise was not only listed as one of many recommendations for general health, but was accorded a unique status in its own right. Further academic treatises on therapeutic riding were to follow,[70] but they all remained attached to theories focusing upon the 'hydraulics' of bodily fluids. That approach was similarly manifest in the medical theory of movement promulgated by Joseph Clément Tissot (1747–1826). He also produced a vernacular (French) treatise, entitled *Gymnastique medicinale* (1780), which contained a dedicated chapter advocating the exercise of moderate riding without too violent agitation.[71]

Therapeutic riding as it is understood today, however, did not appear until after the Second World War, especially in Germany, Austria and Switzerland. It was no longer used so much in cases of lack of movement, constipation and other assumed disorders in the flow of bodily fluids, but increasingly for psychosocial benefits and improvements to the motor senses.[72] It was essential for the whole experience that the horse just could not be simply replaced by an artificial apparatus, because the specific success is based on handling a living creature.

68 Cf. Krause, Ehrendenkmahl, p. 31–32.

69 In the original: "Er war bey jedermann beliebt, gegen alle, besonders gegen die Armen, gütig, welchen letztern er mit seiner Hülfe ohne Entgeld beygestanden, und ihnen die heilsamsten Arzneyen umsonst gegeben hat." (Krause, Ehrendenkmahl, p. 30).

70 See, e.g., Andreas Ramzelius with *De equitatione eiusque in medicina usu* (1738), Christian Michael Adolphi and Carl Friedrich Breitenbach with *De equitationis eximio usu medico* (1744), or Johann Friedrich Wilhelm Dietz and Stephan Zágoni Bara withHau *De salutari limitando tamen equitationis exercitio* (1757).

71 Cf. Joseph Clément Tissot: Gymnastique médicinale et chirurgicale, ou Essai sur l'utilité du movement ou des différens exercices du corps et due repos dans la cure des maladies. Paris 1780, p. 111–119.

72 See as a groundbreaking work for the development of theory in therapeutic horseback riding the congress volume by Hermann Bünte, Heinrich Beck and Max Reichenbach (ed.): Reiten als Therapie. Erlangen, Basel, München [et al.] 1972 (Beiträge zur Sportmedizin Vol. 3).

Against this background, it is finally noteworthy that Quellmaltz, for all the enthusiasm for technology of his age, also very firmly emphasised the limits of his apparatus. On the one hand, he put the therapeutic effect of the riding machine into perspective. He wished to see its use only as an accompanying measure to the parallel use of surgery, pharmacy and dietetics (cf. Q, p. 15 and 47). On the other hand, he considers he considered that riding real horses was much better than riding his *cheval-machine*. He stated that he did not mean to interfere with the "interest of horse lenders" as he would always "rather refer those who can afford it" to undertake "proper riding" (Q, p. 18).[73] Quellmaltz did not explicitly disclose why he preferred proper riding. He knew that this exercise was valued as a mark of status and wealth, since riding, when it was not a means of earning a living, used to be a privilege of the upper classes – a point which he mentioned in his treatise (cf. Q, p. 2–3). But Quellmaltz may also have been thinking of the unique effects of interaction between humans and animals, which cannot be replaced by machines. Contemporaries were undoubtedly aware of the special sensitivity of horses towards human beings, as seen for example, in Pinter:

> But horses love especially those who love them back, whoever they may be, which they do not only recognise easily through many obvious expressions of love, but they also have a much more subtle understanding of the natural counter-love, which is also innate in dogs and other animals, from an inner sympathy.[74]

Nonetheless, while the substitution of the horse by a riding machine was regarded by Quellmaltz as a second-choice option in medical therapy, 'riding machines' were to be found as fitness equipment from the eighteenth century to the present day. Thus artificial horses are still used in some circumstances for training riders in the skills of equitation. One example can be seen in the machines produced by the English company Racewood. These have extremely

73 In the original: "Jnteresse derer Pferdeverleiher", "lieber zum ordentlichen Reuten verweisen" (Q, p. 18).

74 In the original: "Es lieben aber die Pferde absonderlich / die sie wieder lieben / wer dieselbe auch seyn mögen / welches sie auß vielerley offentlichen Liebs Bezeigungen/ nicht allein leichtlich abmercken/ sondern auch auß einer innerlichen Symparia [!] haben sie noch viel ein subtilere Erkantnuß der natürlichen Gegenliebe/ welche auch den Hunden und andern Thieren mehr angeboren ist." (Johann Christoph Pinter: Vollkommener ergäntzter Pferd=SchatzDemo [1st main part]. Frankfurt/M. 1664, p. 88).

complex digital controls, and are by no means devised only for use by beginners.[75] Last but not least, a cultural fascination with mechanical imitations of nature remains unbroken in the performative and visual arts to this day. Large mechanical horses were used as equine 'actors', for instance, when the National Theatre in London produced Nick Stafford's 2007 stage adaptation of Michael Morpurgo's novel *War Horse* (1984) in cooperation with the South African Handspring Puppet Company.[76] One mobile sculpture by Adrian Landon, entitled *Mechanical Horse* (2015), presents another fine example, having no apparent purpose beyond its aesthetic value.[77] The confirmed do-gooder Quellmaltz would no doubt have admired the technical ingenuity but equally frowned upon the absence of tangible social benefits. The aim of his newly-invented 'Pegasus' was to place machines at the service of all humanity.

75 See the demo videos on the company's website at https://www.racewood.com/videos.html [12.09.2020].

76 See the official trailer on the National Theatre's YouTube channel at https://www.youtube.com/watch?v=pdhKmRak2NM [16.06.2021].

77 A video of the moving sculpture is linked to the artist's website: https://www.youtube.com/watch?v=-JURLtuz1f8&feature=emb_title [30.08.2021].

CHAPITRE 6

« Les animaux, nos confrères » dans l'œuvre de Voltaire

Halima Ouanada

Résumé

Cette contribution examine quelques textes de Voltaire, qui témoignent de son intérêt marqué pour la 'question animale' et de sa pertinence au sein de son œuvre, notamment dans sa réflexion sur l'anthropocentrisme. Malgré des points de vue qui semblent au premier abord très clairs et définitifs sur la question animale, l'article souligne les ambiguïtés propres à la pensée de Voltaire. Il met en évidence certaines hésitations de l'auteur des Lumières et s'intéresse aux influences directes et indirectes qu'elles reflètent ainsi qu'aux raisons profondes de sa pensée.

Abstract

This contribution examines some highly topical texts by Voltaire, which show a clear interest in the 'animal question', particularly in reference to anthropocentrism. However, the clarity of Voltaire's approach is shot through with a degree of ambiguity. His mixed responses are accordingly highlighted here, as a means of elucidating the direct and indirect influences upon him – and the deeper reasons behind his thinking.

Nous tentons avec la présente contribution d'examiner quelques textes de Voltaire (1694–1778) qui s'inscrivent fortement dans notre actualité et dans lesquels l'auteur témoigne d'un intérêt certain pour la question animale, notamment dans son rapport avec l'anthropocentrisme[1]. C'est néanmoins un

1 Pour la phrase «Les animaux, nos confrères», voir Lettre de Voltaire (1775). Dans: Idem: Œuvres complètes de Voltaire. Éd. par Theodore Besterman. Oxford 2006 (Vol. 74B); cité par Renan Larue: Le végétarisme dans l'œuvre de Voltaire (1762–1778). Dans: Dix-huitième siècle 42 (2010), no. 1, 19–34, ici p. 21. – Dans l'article 'Viande', Voltaire dira « les animaux nos semblables » (Voltaire: [Art.] 'Viande'. Dans: idem: Questions sur l'Encyclopédie. Œuvres complètes de Voltaire. Éd. par Louis Moland. Vol. 20: Dictionnaire philosophique. Vol. 4. Paris 1879, p. 577).

 | DOI:10.1163/9789004495395_007

intérêt entaché d'une ambiguïté que nous jugeons utile de mettre en lumière afin d'en dégager les influences et les raisons directes et indirectes. En effet, loin d'être une prise de conscience toute récente, fruit du progrès scientifique et de l'influence des organisations pour la défense des droits des animaux, la question animale – les liens qui unissent les animaux aux hommes, leur sensibilité propre comme la part de responsabilité que les hommes prennent à leurs souffrances – a beaucoup préoccupé les esprits du siècle des Lumières. Un simple détour par les célèbres tableaux de Jean Siméon Chardin (1699–1779)[2], ou par les livres de cuisine de l'époque rendent aisément compte de cet intérêt recrudescent pour l'animal et des débats qu'il suscite[3].

Voltaire, en digne représentant de son siècle, consacre à cette question animale un certain nombre de textes virulents qu'on retrouve aisément dans les *Questions sur l'Encyclopédie* (4 vols., 1770–74) et dans le *Dictionnaire philosophique* (1764). Ce sont des textes qui pourraient servir aujourd'hui de fer de lance pour la défense et la protection des animaux. Cette citation en est un exemple: « Il ne leur manque que la parole; s'ils l'avaient, oserions-nous les tuer et les manger? Oserions-nous commettre ces *fratricides*?[4] Quel est le barbare qui pourrait faire rôtir un agneau, si cet agneau nous conjurait par un discours attendrissant de n'être point à la fois assassin et anthropophage?[5] »

Fortement influencé par Porphyre dont il vante la sagesse et la justice envers les animaux, mais aussi par Plutarque, connu pour son végétarisme et auteur de *L'intelligence des animaux*, comme par le philosophe anglais John Locke, auteur de l'*Essai sur l'entendement humain* (1689), Voltaire semble puiser, chez

---

2 « Chardin réalise souvent des séries, mettant en scène les mêmes animaux et objets dans des situations et des postures très comparables. L'animal y est le plus fréquemment mort et se trouve en général, non pas dans l'espace prestigieux et aristocratique de la chasse, comme le font fréquemment les peintres animaliers officiels, mais dans celui, clos et banal, de la cuisine où il est préparé pour être consommé, suivant en cela une tradition hollandaise. » (Jean-Luc Guichet: L'animal dans la peinture de Chardin. Dans: Dix-huitième siècle 36 (2004), p. 547–556, ici p. 547).

3 Voir Beatrice Fink: Introduction. Dans: eadem (dir.): Les Liaisons savoureuses. Réflexions et pratiques culinaires au XVIIIe siècle. Saint-Etienne 1995 (Lire le Dix-huitième Siècle), p. 7– 20, ici p. 7; « [de] 1730 à 1770, titres, éditions se multiplient à une cadence sans précédent, croissance retombant rapidement dans les deux décennies qui précèdent la Révolution. La concurrence entre les éditeurs et leurs auteurs est vive. Entre 1739 et 1759 sont mis en circulation 56,5 % de tous les titres produits entre 1700–1789. Période d'invention culinaire – une 'nouvelle cuisine' s'élabore – où les goûts changent, se diffusent dans un public élargi » (Alain Girard: Le Triomphe de 'la cuisinière bourgeoise'. Livres culinaires, cuisine et société en France aux XVIIe et XVIIIe siècles. Dans: Revue d'histoire moderne et contemporaine 24 (1977), no. 4, p. 498–523, ici p. 508.

4 C'est nous qui mettons en italiques.

5 Voltaire: [Art.] 'Viande', Questions sur l'Encyclopédie, p. 577.

ces auteurs antiques et modernes, les arguments pour condamner la théorie cartésienne de l'animal-machine, critiquer la doctrine chrétienne et réfléchir à sa manière à l'origine du mal: « Quelle pitié, quelle pauvreté, d'avoir dit que les bêtes sont des machines privées de connaissance et de sentiment », écrit-il en tête de son article 'Bêtes' du *Dictionnaire philosophique*[6].

Rappelons rapidement que la question du rapport de l'homme à l'animal n'est tout de même pas une invention du siècle des Lumières. Il suffit de parcourir les écrits des auteurs de l'Antiquité grecque et de s'arrêter quelque peu sur la production du seul Aristote, par exemple, pour se rendre compte de l'ancienneté de cette problématique[7], comme du débat qu'elle suscite.

Par ailleurs, si l'intérêt porté à la question animale s'affaiblit au Moyen Âge, c'est à la Renaissance, surtout avec Michel de Montaigne (1533–1592), qu'il se trouve remis à l'ordre du jour. Ce dernier reprend à son compte le débat pour fustiger le rationalisme anthropocentriste d'Aristote dans une tentative de mettre l'homme sur un pied d'égalité d'avec l'animal[8]. En effet, s'inscrivant dans le sillage de Plutarque, Montaigne considère qu' « il y a plus de distance de tel homme à tel homme qu'il n'y a de tel homme à telle bête »[9] Pour lui, les animaux sont « nos confrères et compagnons »[10].

Toutefois, cette théorie ne tardera pas à être remise en question par la conception de l'animal-machine défendue par René Descartes au XVIIème siècle, reprise et poussée à l'extrême par les postcartésiens, notamment Nicolas Malebranche (1638–1715). Dépourvu d'âme et de sensibilité, l'animal, selon René Descartes, est une pure machine. L'idée se répand durant toute la deuxième moitié du XVIIème siècle et fournit même des « *topoï* à la littérature »[11]. Elle fait toutefois débat et des voix incrédules s'élèvent parmi elles celle d'une femme, Madame de Sévigné, pour déclarer: « Des machines qui

6 Voltaire: Dictionnaire philosophique. Éd. par Raymond Naves et Olivier Ferret. Paris 2008, p. 50–53.

7 A lui seul, Aristote a produit un ensemble de traités: l'*Histoire des animaux* (*Peri ta zoa historiai*), les *Parties des animaux* (*Peri zoon morion*), la *Génération des animaux* (*Peri zoon geneseos*), le *Mouvement des animaux* (*Peri zoon kineseos*), qui a inspiré, entre autres – Pline l'Ancien qui a consacré quatre livres de son *Histoire naturelle* aux animaux.

8 Voir Michel de Montaigne: Essais. Vol. 2. Éd. par Pierre Villey. 2ème éd., Paris 1992 (Quadrige Vol. 94), p. 481.

9 Ibid., p. 481.

10 Ibid.

11 Cf. Delphine Reguig-Naya: Descartes à la lettre: poétique épistolaire et philosophie mondaine chez Mme de Sévigné. Dans : Dix-septième siècle 216 (2002), no. 3, p. 511–525; en ligne: https://www.cairn-int.info/revue-dix-septieme-siecle-2002-3-page-511.htm [06.11.2020].

aiment, des machines qui ont une élection pour quelqu'un, des machines qui sont jalouses, des machines qui craignent. Allez, allez-vous vous moquer de nous: jamais Descartes n'a prétendu nous le faire croire »[12].

Au siècle des Lumières, c'est le patriarche de Ferney qui s'inscrit en faux contre cette idée du dualisme de l'âme et du corps de Descartes, suivant laquelle les bêtes sont semblables à des horloges et donc dépourvues de sensibilité. Tout comme Locke, dont il s'inspire, Voltaire condamne la théorie cartésienne et reconnaît à l'animal une sensibilité comparable à celle de l'homme. Son article 'Bêtes' du *Dictionnaire philosophique* constitue déjà une réponse à la théorie de l'animal-machine développée par Descartes en 1646. Dans cet article, Voltaire commence par établir un parallèle entre la sensibilité humaine et la sensibilité animale :

> Est-ce parce que je te parle que tu juges que j'ai du sentiment, de la mémoire, des idées ? Eh bien ! je ne te parle pas ; tu me vois entrer chez moi l'air affligé, chercher un papier avec inquiétude, ouvrir le bureau où je me souviens de l'avoir enfermé, le trouver, le lire avec joie. Tu juges que j'ai éprouvé le sentiment de l'affliction et celui du plaisir, que j'ai de la mémoire et de la connaissance. Porte donc le même jugement sur ce chien qui a perdu son maître, qui l'a cherché dans tous les chemins avec des cris douloureux, qui entre dans la maison, agité, inquiet, qui descend, qui monte, qui va de chambre en chambre, qui trouve enfin dans son cabinet le maître qu'il aime, et qui lui témoigne sa joie par la douceur de ses cris, par ses sauts, par ses caresses[13].

L'auteur attire ensuite l'attention du lecteur sur les conséquences désastreuses de cette hypothèse pour l'espèce animale. Il en profite pour dénoncer le caractère violent des expériences scientifiques faites, à l'époque, sur les animaux :

> Des barbares saisissent ce chien, qui l'emporte si prodigieusement sur l'homme en amitié ; ils le clouent sur une table, et ils le dissèquent vivant pour te montrer les veines mésaraïques. Tu découvres dans lui tous les mêmes organes de sentiment qui sont dans toi. Réponds-moi, machiniste, la nature a-t-elle arrangé tous les ressorts du sentiment dans cet animal, afin qu'il ne sente pas ? A-t-il des nerfs pour être impassible ? Ne suppose point cette impertinente contradiction dans la nature[14].

---

12 Madame de Sévigné, citée par Jacques Debû-Bridel: Un vieil homme et son caniche. Paris 1980, p. 65.

13 Voltaire: [Art.] 'Bêtes', Dictionnaire philosophique, p. 51–52.

14 Ibid., p. 52.

Il affiche enfin ouvertement son dégoût pour le spectacle de la boucherie, loue la tempérance des brahmanes, « *les premiers qui s'imposèrent la loi de ne manger d'aucun animal* »[15] et appelle carrément à condamner ses pratiques :

> Il n'est que trop certain que ce carnage dégoûtant, étalé sans cesse dans nos boucheries et dans nos cuisines, ne nous paraît pas un mal, au contraire, nous regardons cette horreur, souvent pestilentielle, comme une bénédiction du Seigneur et nous avons encore des prières dans lesquelles on le remercie de ces meurtres. Qu'y a-t-il pourtant de plus abominable que de se nourrir continuellement de cadavres[16].

Devenu un personnage littéraire sous la plume de Voltaire, l'animal est fortement présent dans son œuvre où il prend la parole pour faire part de son supplice et pour décrire l'extrême violence subie de la part des humains. L'exemple le plus parlant se trouve sans doute dans le dialogue du *Chapon et de la Poularde* (1763) qui décrit de manière édifiante les atrocités endurées par les bêtes pour satisfaire les papilles des gourmands. Ce dialogue composé avant même l'article 'Bêtes' du *Dictionnaire philosophique*, attire l'attention sur la ressemblance entre l'humain et l'animal quant à la sensibilité. D'où s'en suit un appel à arrêter le massacre. Jean-Jacques Rousseau ira dans le même sens, car pour lui, la souffrance de l'animal est une preuve de sa sensibilité et de sa ressemblance avec l'humain. En 1755, il écrit :

> Si je suis obligé de ne faire aucun mal à mon semblable, déclare-t-il, en effet, c'est moins parce qu'il est un être raisonnable que parce qu'il est un être sensible ; qualité qui, étant commune à la bête et à l'homme, doit au moins donner à l'une le droit de n'être point maltraitée inutilement par l'autre[17].

Pour Voltaire, il n'existe aucune différence de nature entre les êtres humains et les animaux, comme il l'expose déjà dans son *Traité de métaphysique* (1734), qui contient des discussions pertinentes sur la rationalité et les compétences des animaux par rapport aux êtres humains.

---

15 Voltaire: [Art.] 'Viande', Questions sur l'Encyclopédie, p. 577.

16 Voltaire: Il faut prendre un parti, ou Le principe d'action. Dans: idem: Œuvres complètes de Voltaire. Éd. par Theodore Besterman. Oxford 2006 (Vol. 74B), p. 38.

17 Jean Jacques Rousseau: Discours sur l'origine et les fondements de l'inégalité parmi les hommes. Éd. par Jean Starobinski. Paris 1965, p. 55–56.

La seule différence est certainement de degré, en référence à la célèbre allégorie de la grue[18], chère à Platon. C'est une idée que l'on retrouve également chez Denis Diderot, notamment dans son article 'Animal' de l'*Encyclopédie ou Dictionnaire raisonné des sciences, des arts et des métiers* (1751–80), où il interpelle à ce sujet Georges-Louis Leclerc de Buffon, l'auteur de l'*Histoire naturelle* (1749–67)[19], dans les termes suivants: « l'expérience marchant pied à pied, et l'esprit de système allant toujours par sauts et par bonds »[20]. En effet, cette hiérarchie entre les espèces et cette supériorité de l'homme sur l'animal, prêchée par Buffon et par la mouvance religieuse notamment, n'est, pour Voltaire, que chimère et pur produit des fables des théologiens. Car, à observer les animaux, rien ne les distingue des humains. Ils sont même par certains aspects plus organisés et bien supérieurs. Il interroge en ce sens :

> [...] quel est le chien de chasse, l'orang-outang, l'éléphant bien organisé qui n'est pas supérieur à nos imbéciles que nous renfermons, à nos vieux gourmands frappés d'apoplexie, traînant les restes d'une inutile vie dans l'abrutissement d'une végétation interrompue, sans mémoire, sans idées, languissant entre quelques sensations et le néant ? Quel est l'animal qui ne soit pas cent fois au-dessus de nos enfants nouveau-nés ?[21]

Être humain n'est donc pas un gage de supériorité pour Voltaire, mais devrait davantage pousser à adopter une posture d'humilité sur la question animale.

---

18 « L'allégorie de la grue a pour fonction de montrer que la différence entre l'homme et l'animal varie selon le degré d'âme. L'homme et l'animal sont caractérisés par des morphologies psychiques [...]: l'homme a la raison en partage, l'animal est dominé par son désir, puisqu'il n'a pas la raison en partage, etc. Mais de nouveau, certains hommes peuvent être des bêtes, lorsque le désir détermine leur comportement. En ce sens, l'animalité serait davantage un critère éthique qu'un critère naturel déterminé ». (Mathieu Hilfiger: 'L'humanité' chez Platon. Dans: Le Philosophoire 23 (2004), no. 2, p. 175; en ligne: https://www.cairn.info/revue-le-philosophoire-2004-2-page-166.htm [21.12.2020]).

19 Voir Georges-Louis Leclerc Comte de Buffon: Histoire naturelle. Éd. par Jean Valoot. Paris 1984 (Collection folio Vol. 1569), p. 204.

20 Il faut dire que, pour Buffon, l'humain l'emporte sur l'animal en raison justement du pouvoir de l'esprit, apanage de la seule espèce humaine, en plus de son homogénéité biologique, à l'origine de la création: « l'être le plus noble de la création, l'espèce en est unique, puisque les hommes de toutes les races [...] peuvent se mêler et produire ensemble. » Buffon: [Art.] 'Lion'. Dans: idem : Histoire naturelle, p. 204.

21 Voltaire: Les Adorateurs ou les louanges de Dieu. Ouvrage unique de M. Imhof, traduit du Latin (1769). Dans: idem: Œuvres complètes de Voltaire. Éd. par Louis Moland. Vol. 28, p. 309–326, ici p. 319. – Cf. Renan Larue: Le végétarisme dans l'œuvre de Voltaire (1762–1778). Dans: Dix-huitième siècle 42 (2010), no. 1 [numéro spécial: L'animal des Lumières, ed. par Jacques Berchtold et Jean-Luc Guichet], p. 19–34, ici p. 23.

Ainsi, pour le patriarche de Ferney, comme pour bon nombre de ses contemporains, l'éthique se fonde sur la conscience d'une sensibilité en commun avec les animaux envers lesquels nous, les hommes, devons sentir plutôt de la pitié. Défenseur de la conception matérialiste de l'être humain, Julien Offray de La Mettrie (1709–1751), un matérialiste radical, par exemple, va jusqu'à convoquer l'anatomie comparée pour prouver la ressemblance entre homme et animal, comme le fait, par exemple, Boerhaave[22]. Il suit en cela Locke qui soutient « que Dieu, qui a donné aux bêtes la faculté de s'apercevoir, de se souvenir, d'avoir quelques idées, [a] pu communiquer à nos organes plus déliés une intelligence bien supérieure »[23]. Pour David Hume, la ressemblance intellectuelle entre l'homme et la bête est incontestable. « Aucune vérité », dit-il, « ne me paraît plus évidente que de dire que les bêtes sont douées de pensée et de raison tout comme les hommes »[24]. Seul le langage fait la différence pour Rousseau qui, dans son *Discours sur l'origine et les fondements de l'inégalité parmi les hommes* de 1755, fait écho à la citation de Voltaire selon laquelle « il ne leur manque que la parole »[25]. Toutefois, on est en droit de se demander si la parole allait de pair avec les actes, chez Voltaire? Etait-il devenu végétarien pour autant? Des recherches actuelles semblent plutôt indiquer le contraire[26].

La réflexion menée par Voltaire sur l'animal est entachée d'ambiguïté, comme d'ailleurs la majorité des questions les plus brûlantes du siècle des Lumières, telles celle de l'esclavage, du colonialisme et du rôle des femmes dans la société. La question animale est, elle aussi, sujette à une inconséquence notoire de la part de l'auteur dont les prises de position ne correspondent pas toujours à ses pratiques. Comme le précise Christiane Mervaud, dans son ouvrage *Voltaire à table: Plaisir du corps, plaisir de l'esprit*[27], les livres

22 Cf. Theodorus H. M. Verbeek: Traité de l'âme de La Mettrie. Utrecht [diss.] 1988, p. 22–23.

23 Cf. Herman Boerhaave: Institutions de médecine de M. Hermann Boerhaave. 2ème éd., Paris 1743, p. 104. – Voir Ann Thomson: La Mettrie, lecteur et traducteur de Boerhaave. Dans: Dix-huitième siècle 23 (1991), p. 23–29; Ann Thomson: Les « Animaux plus que machines ». Dans: Dix-huitième siècle 42 (2010), no. 1 [numéro spécial: L'animal des Lumières, ed. par Jacques Berchtold et Jean-Luc Guichet], p. 339–351; en ligne: https://www.cairn.info/revue-dix-huitieme-siecle-2010-1-page-339.htm [13.11.2020].

24 David Hume: L'Entendement. Traité de la nature humaine. Livre I et appendice (1739). Éd. par Philippe Baranger et Philippe Saltel. Paris 1995, p. 254 (t. III, section XVI).

25 Voltaire: [Art.] 'Viande', Questions sur l'Encyclopédie, Œuvres complètes de Voltaire, p. 577.

26 A propos de ce débat voir Christiane Mervaud: Voltaire à table. Plaisir du corps, plaisir de l'esprit. Paris 1998 (Le bon sens), p. 161; eadem: Bestiaires de Voltaire. Oxford 2006 (Studies on Voltaire and the Eighteenth Century Vol. 6), p. 49–54.

27 Mervaud, Voltaire à table, p. 48 et p. 56. – Cf. Larue, Le végétarisme dans l'œuvre de Voltaire (1762–1778), p. 20.

de compte de Ferney révèlent des quantités considérables de viandes recherchées (dindons, canards, lapins, ortolans ...) que le patriarche de Ferney offrait à profusion à ses convives et ce au moment même où il écrivait le dialogue critique des cruautés envers les animaux du *Chapon et la Poularde*[28]. L'auteure rappelle, de surcroît, qu'il se faisait scrupule de manger les huitres crues, ce qui ne l'empêchait pas pour autant de réaliser des expériences de décapitation sur les escargots[29]. Le comportement inconséquent de Voltaire, était « à l'image d'un certain nombre de ses contemporains, touchés de la mort des animaux de boucherie mais peu enclins à se passer de viande » résume Renan Larue[30].

Rappelons encore que les prises de position fermes, parfois même emportées, en faveur des animaux n'ont pas toujours été de mise dans son œuvre. Bien au contraire, avant l'année 1762, le patriarche de Ferney tenait un discours bien différent où la bonne cuisine, notamment les mets carnés, relevait chez lui d'une forme d'esthétique et désignait des plaisirs dont il fallait absolument pouvoir jouir, comme il l'explique dans son article 'Goût' :

> Ce sens, ce don de discerner nos aliments, a produit dans toutes les langues connues, la métaphore qui exprime par le mot goût, le sentiment des beautés et des défauts dans tous les arts: c'est un discernement prompt comme celui de la langue et du palais, et qui prévient comme lui la réflexion ; il est comme lui sensible et voluptueux, à l'égard du bon, il rejette comme lui le mauvais avec soulèvement ; il est souvent, comme lui, incertain et égaré, ignorant même si ce qu'on lui présente doit lui plaire, et ayant quelquefois besoin comme lui d'habitude pour se former[31].

Il ajoute encore dans ce même article: « Il ne suffit pas pour le goût, de voir, de connaître la beauté d'un ouvrage; il faut la sentir, en être touché »[32]. C'est justement, ce caractère prestigieux du repas, reflet d'une certaine abondance et d'un raffinement opulent chers au siècle et à Voltaire, qui aurait été à l'origine

---

28 Cf. Mervaud, Voltaire à table, p. 48 et p. 56.

29 Mervaud, Bestiaires de Voltaire, p. 39, et Larue, Le végetarisme dans l'œuvre de Voltaire (1762–1778), p. 22.

30 Larue, Le végetarisme dans l'œuvre de Voltaire (1762–1778), p. 22.

31 Voltaire: [Art.] 'Goût'. Dans: Encyclopédie ou Dictionnaire raisonné des sciences, des arts et des métiers. [Sous la direction de Denis Diderot et Jean-Baptiste le Rond D'Alembert]. Lausanne et Berne 1780–1782, t. XVI, p. 336. – Voir Maddalena Mazzocut-Mis: « La bonne cuisine et le siècle des Lumières ». Nouvelle revue d'esthétique 14 (2014), n° 2, p. 115–129; en ligne: DOI: 10.3917/.014.0115; https://www.cairn-int.info/revue-nouvelle-revue-d-esthetique-2014-2-page-115.htm [06.11.2020].

32 Voltaire: [Art.] 'Goût', p. 336.

de la célèbre maladie de la goutte, appelée également la reine des maladies, car elle était considérée comme la maladie des bons vivants.

Par ailleurs, dans le débat du siècle sur le progrès culinaire et l'innovation gastronomique, dont l'article 'Cuisine' de Jaucourt pour l'*Encyclopédie* fait l'écho, Voltaire s'est rangé dans le camp des bons vivants conservateurs contre celui des progressistes[33]. Il déclare à son ami d'Argental, à propos de la nouvelle cuisine moderne à la mode: « Je ne puis manger d'un hachis composé de dinde, de lièvre et de lapin, qu'on veut me faire prendre pour une seule viande. Je n'aime ni le pigeon à la crapaudine, ni le pain qui n'a pas de croûte. Je bois du vin modérément, et je trouve fort étranges les gens qui mangent sans boire, et qui ne savent pas ce qu'ils mangent »[34]. Il vante encore la cuisine traditionnelle dont il fait la description dans le festin offert aux trois prophètes dans le chapitre VI du *Taureau blanc*, paru en 1774, quatre ans avant sa mort :

> On servit deux potages, l'un de bisque, l'autre à la reine ; les entrées furent une tourte de langues de carpes, des foies de lottes et de brochets, des poulets aux pistaches, des innocents aux truffes et aux olives, deux dindonneaux au coulis d'écrevisses, des mousserons et des morilles, et une chipolata. Le rôti fut composé de faisandeaux, de perdreaux, de gelinottes, de cailles et d'ortolans, avec quatre salades. Au milieu était un surtout dans le dernier goût. Rien ne fut plus délicat que l'entremets ; rien de plus magnifique, de plus brillant et de plus ingénieux que le dessert[35].

Voltaire a-t-il vraiment changé d'avis et a-t-il arrêté d'aimer la cuisine carnée? Faut-il le croire sur parole quand il déclare, à soixante-huit ans: *« Je ne mange plus de viande »?*[36]

---

33 « La cuisine simple dans les premiers âges du monde, devenue plus composée et plus raffinée de siècle en siècle [...] est actuellement une étude, une science des plus pénibles, sur laquelle nous voyons paraître sans cesse de nouveaux traités sous les noms de Cuisinier François, Cuisinier royal, Cuisinier moderne, Dons de Comus ou l'Art de la cuisine, École parfaite des officiers de bouche, et beaucoup d'autres qui changeant perpétuellement de méthode, prouvent assez qu'il est impossible de réduire à un ordre fixe, ce que le caprice des hommes et le dérèglement de leur goût, recherchent, inventent, imaginent, pour masquer les aliments. » François de Jaucourt: [Art.] 'Cuisine'. Dans: Encyclopédie. Vol. X., p. 107–108.

34 Voltaire: À M. le Comte d'Argental, 4 septembre 1765. Dans: Œuvres complètes de Voltaire. Vol. XLVII. Paris 1830, p. 429.

35 Voltaire: Le Taureau blanc [1774]. Dans: Œuvres complètes, vol. XXI, Paris 1877, p. 504–505.

36 Voir les Lettres D15565, D21001 et D. 21162 (lettre du 22 avril 1778): « Comment puis-je être empoisonné par tant de glaire dans les entrailles, lorsque je ne mange ni viande ni

En réalité, on voit mal l'auteur du poème *Le Mondain* (1736), célèbre apologète du luxe, connu pour ses réceptions, prêcher le végétarisme et le rejet de la consommation des animaux qui constituent la base de toute cuisine qui se respecte à l'époque, et de la sienne en particulier[37]. « Voltaire, lorsqu'il n'était pas trop souffrant, explique Jean Orieux, faisait honneur à ce luxe de la table auquel il tenait »[38].

Deux explications sous-tendent, nous semble-t-il, son attitude vis-à-vis des mets carnés, l'une physique et l'autre idéologique. La première est sans doute relative à l'état de santé du patriarche. Ayant perdu très tôt ses dents et souffrant constamment de coliques et de la goutte entre autres maladies, le patriarche de Ferney dont la *Correspondance* regorge de plaintes, se trouve obligé d'adopter une autre posture par rapport à l'animal. Dans son compte-rendu d'un ouvrage biographique de Voltaire, Pierre Julien note à ce propos :

> Tout homme d'esprit qu'il était, Voltaire, ne fut-il pas aussi un malade imaginaire ? La quantité de médicaments qu'il absorba dépasse l'imagination. Il a pour sa santé des remparts de pilules ... Il a plus de mille fioles ... Il se procure tous les remèdes nouveaux et les ingurgite au plus vite ... Il souffre de la colique ... Il est deux jours au lit ... Il est mourant ... Il a la fièvre tierce ... Il a la fièvre quarte ... Il est persuadé qu'il mourra avant quarante ans. Puis un jour il découvre les pilules de Staal. Elles sont merveilleuses. Il en prend des quantités à faire mourir un homme robuste. Le Dr Purgon n'eut jamais meilleur malade. Quand il a la petite vérole, on lui fait absorber huit doses d'émétique et deux cents pintes de limonade. Il lui arrive de prendre en un mois huit purges et douze lavements. Il se plaisait à répéter que la médecine consiste à introduire des drogues que l'on ne connaît pas dans un corps que l'on connaît encore moins. Et immédiatement il recourait non seulement aux médecins classiques, au baume tranquille, à l'eau de Rabel, au baume de Varenger, mais aussi aux remèdes de bonne femme, à la grenaille de fer, au petit lait, à l'essence de cannelle. Voltaire craignit toujours que son médecin ne le crût en bonne santé. Lui dire qu'il se portait bien, c'était le calomnier. Le féliciter sur sa

poisson? », écrit Voltaire à Tronchin. – Cf. Larue, Le végétarisme dans l'œuvre de Voltaire (1762–1778), p. 21.

37 Sur le débat à propos de l'hypothèse d'un végétarisme voltairien. – Cf. Renato Galliani: Voltaire, Porphyre et les animaux. Dans: Studies on Voltaire [dorénavant SVEC] 199 (1981), p. 125–138. – Voir aussi Mervaud, Voltaire à table, p. 161; Mervaud, Bestiaires de Voltaire, p. 49–54.

38 Jean Orieux: Voltaire I. Paris 1977, p. 240.

> mine, c'est l'injurier. Pour lui plaire, il faut s'effrayer de sa maigreur, de sa faiblesse, il faut le faire plus vieux et plus ravagé qu'il n'est[39].

Deuxièmement, il se trouve que la question animale, se prête aisément à son combat contre l'Infâme, ce terme de ralliement de son combat contre les abus du pouvoir clérical. Elle permet de contester l'idée de l'âme humaine, avancée par les prêtres qui est, pour lui, un pur produit de l'orgueil humain et un prétexte pour asseoir la supériorité humaine sur la nature.

> À un âge où la Terre a cessé d'être le centre du cosmos, ne devrions-nous pas abandonner notre fantasme d'élection métaphysique et dissiper définitivement l'illusion anthropocentriste entretenue par notre vanité ? L'âme spirituelle, cette « idée creuse » que nous chérissons tant, n'est aux yeux de Voltaire qu'une invention de théologiens qui trouvèrent là un moyen commode de nous distinguer des autres animaux et de justifier leur asservissement[40].

Relevant davantage d'une « tactique d'accaparement »[41], l'animal permet à Voltaire comme le suggère Renan Larue, « d'envisager sous un angle neuf l'origine du mal, ou la validité de l'anthropocentrisme. L'évocation, voire la défense, du végétarisme servira aussi très souvent la cause de l'anticléricalisme »[42]. Pivot d'une philosophie matérialiste pour certains ou du déisme dans le cas de Voltaire, la question animale suscite l'intérêt donc, davantage pour son utilité dans le combat philosophique et politique mené contre l'Infâme que pour elle-même.

C'est certainement pour cette raison que Voltaire, comme Diderot ou La Mettrie, reconnaît la sensibilité des animaux, mais s'accommode parfaitement des expériences dont ils font l'objet. Comme il s'accommode également des phénomènes et des pratiques de l'époque telle que la 'Bestialité', définie dans l'*Encyclopédie* comme « le crime d'un homme ou d'une femme qui engagerait

39 Pierre Julien: Les maladies et la mort de Voltaire. Compte-rendu de Jean Orieux, Voltaire ou la Royauté de l'esprit. Dans: Revue d'histoire de la pharmacie 58 (1970), no. 204, p. 60–61, ici p. 60.

40 Renan Larue: Introduction à Voltaire. Pensées végétariennes. Paris 2014 (Mille et une nuits Vol. 632), p. 57.

41 Larue, Le végétarisme dans l'œuvre de Voltaire (1762–1778), p. 19.

42 Ibid., p. 19.

un commerce charnel avec une bête. Ce crime se punit par le feu: on brûle même l'animal qui a été l'instrument du crime »[43].

L'œuvre monumentale de Voltaire, écrit Vincent Jolivet, se révèle tout à fait emblématique de ce dix-huitième siècle qui s'est aussi intéressé à la question des accouplements entre hommes et bêtes. Voltaire revient sur le sujet à de très nombreuses reprises tout au long de sa carrière, le traitant sous différents angles, aussi bien d'une manière historico-polémique que dans le cadre de ses œuvres fictionnelles[44].

Comme le rappelle Christiane Mervaud, « le philosophe, le moraliste ou l'historien condamnent cette aberration; mais la fable en a rapporté maints exemples et Voltaire se plaît à jouer avec ce sujet ambigu »[45]. Justement, c'est cette ambiguïté qui devient la 'marque de fabrique' de ce philosophe qui hésite souvent à se prononcer clairement à propos de sujets dont il peut tirer profit. En cela, les sources antiques, notamment mythologiques, l'ont déjà initié et ont souvent constitué le socle à partir duquel il réfléchit[46]. Cette ambiguïté enregistrée quant à la question animale, se remarque également quant à la question de la femme, et se révèle donc représentative de celle de la plupart des auteurs du temps qui l'exploitent pour leurs différents engagements dans les querelles et les controverses, mais se dérobent quand il est question d'en assumer les conséquences pratiques. « Indissociable du questionnement sur l'homme, le jugement sur l'animal, note à juste titre Jacques Berchtold, est

43 François-Vincent Toussaint: [Art.] 'Bestialité'. Dans: Encyclopédie ou Dictionnaire raisonné des sciences, des arts et des métiers. [Sous la direction de Denis Diderot et Jean-Baptiste le Rond D'Alembert]. 17 Vols. Paris 1751–1765. Cité par Vincent Jolivet: Lumières et bestialité. Dans: Dix-huitième siècle 42 (2010), no. 1 [numéro spécial: L'animal des Lumières, ed. par Jacques Berchtold et Jean-Luc Guichet], p. 285–303, ici p. 285; en ligne: https://www.cairn.info/revue-dix-huitieme-siecle-2010-1-page-285.htm [06.11.2020].

44 Voir Jolivet, Lumières et bestialité, p. 286–303.

45 Mervaud, Bestiaires de Voltaire, p. 42 et p. 175–180.

46 Voir notamment l'article de Voltaire: 'Bouc, bestialité, sorcellerie'. Dans: Questions sur l'Encyclopédie. Œuvres complètes de Voltaire. Éd. par Theodore Besterman. Vol. 39. Oxford 2008, p. 437: « Mais les boucs reçurent un honneur plus singulier; il est constant qu'en Egypte plusieurs femmes donnèrent avec les boucs le même exemple que donna Pasiphaé avec son taureau. Hérodote raconte que lorsqu'il était en Egypte, une femme eut publiquement ce commerce abominable dans le nome de Mendes: il dit qu'il en fut très étonné, mais il ne dit point que la femme fut punie. Ce qui est encore plus étrange, c'est que Plutarque et Pindare, qui vivaient dans des siècles si éloignés l'un de l'autre, s'accordent tous deux à dire qu'on présentait des femmes au bouc consacré. Cela fait frémir la nature. Pindare dit, ou bien on lui fait dire: Charmantes filles de Mendes,/ Quels amants cueillent sur vos lèvres/ Les doux baisers que je prendrais?/ Quoi! ce sont les maris des chèvres! » – Voir Mervaud, Bestiaires de Voltaire, p. 288.

révélateur de leur audace, mais aussi de leur grande diversité, de leurs ambiguïtés, de leurs contradictions et de leurs tâches aveugles »[47].

Force est de constater donc que comme nombre de questions brûlantes de l'époque, telle que celle du colonialisme, de l'esclavage et de la femme, celle de l'animal ne suscite l'intérêt que pour ce qu'elle offre comme moyen de servir l'idéal de la raison et de combattre l'Infâme. Ce qui expliquerait l'absence de véritable remise en question, de condamnation franche ou d'appel à un vrai changement de comportement face aux animaux. Le patriarche de Ferney ne manifeste de l'intérêt pour l'animal, dont il admet pourtant et la sensibilité et la ressemblance avec l'homme, que pour ce qu'il pourrait apporter comme enseignement sur la nature humaine.

Nul doute de l'amour que le patriarche porte aux animaux domestiques ou de la sincérité de son appel en faveur du végétarisme, du moins en certaines périodes de sa vie, ce qui ne l'empêche pas toutefois d'éprouver du plaisir à déguster des viandes sur une table bien garnie. N'admet-il pas volontiers, en évoquant Plotin, Jamblique et Porphyre, que chez les philosophes, il est « assez rare de faire ce qu'on prêche »[48].

Ainsi, la reconnaissance de la sensibilité, chez les « animaux, nos confrères » n'empêche nullement leur exploitation même de la façon la plus ignoble. De la même manière, se nourrir de la chair humaine, thème qu'on retrouve sous sa plume, relève certes d'une certaine inhumanité, mais la faim et l'instinct de survie pourraient aisément la justifier et même la faire accepter[49]. En somme, la question animale, comme celle de la bestialité, voire celle du cannibalisme, s'inscrivent toutes de plain-pied dans une question centrale aux Lumières, à savoir, « qu'est-ce que l'homme »?[50] Et l'animal n'est qu'une énième « machine à faire penser l'homme »?[51] En ce sens, Diderot n'écrit-il pas à l'article 'Encyclopédie' que « l'homme est le terme unique d'où il faut partir et auquel il faut tout ramener »[52]. Encore faut-il préciser de quel homme il était question.

---

47 Jacques Berchtold et Jean-Luc Guichet: Introduction. Dans: Dix-huitième siècle 42 (2010), no. 1 [numéro spécial: L'animal des Lumières, ed. par Jacques Berchtold et Jean-Luc Guichet], p. 3–18; en ligne: https://www.cairn.info/revue-dix-huitieme-siecle-2010-1-page-3.htm [06.11.2020].

48 Voltaire, Dictionnaire philosophique, Vol. 66, p. 155.

49 Sur la question du cannibalisme, cf. Mervaud, Voltaire à table, p. 166–168.

50 Jolivet, Lumières et bestialité, p. 303.

51 Anne Coudreuse: Le Bestiaire dans *Zadig*. Dans: Revue des sciences humaines (2009), no. 296, p. 117–129, p. 9.

52 Denis Diderot: [Art.] 'Encyclopédie'. Dans: Encyclopédie, Encyclopédie ou Dictionnaire raisonné des sciences, des arts et des métiers. [Sous la direction de Denis Diderot et Jean-Baptiste le Rond D'Alembert.] 17 Vols. Paris 1751–1765. Vol. III, p. 128.

CHAPTER 7

On the Popularity of Songbirds in Eighteenth-Century German Fables

Kristin Eichhorn

Résumé

Dans les pays germanophones, la fable de l'oiseau chanteur a connu une extrême popularité au cours du XVIIIe siècle. Cet article examine ce phénomène dans son contexte esthétique et philosophique. Les oiseaux chanteurs produisant un art qui peut être comparé à la production artistique humaine, leur évocation nourrit ainsi une réflexion sur une forme idéale de coexistence entre les humains au sein d'une société éclairée.

Abstract

In the German-speaking countries, the fable of the songbird became extremely popular during the eighteenth century. This contribution examines this phenomenon in both its aesthetic and its philosophical context. As songbirds produce an art-form that can be compared with human artistic production, the eighteenth-century evocations of songbirds prompted fresh thoughts about an ideal form of human compatibility with other creative species, in an enlightened society.

The fable as a literary genre is known for a wide range of animal protagonists, ranging from wild animals such as the wolf and fox to domesticated animals such as the pig, horse, or cow. Especially in German-speaking countries, however, the eighteenth-century fable at its production peak displayed a striking preference for another species, which has received significantly – yet undeservedly – much less scholarly attention. Yet it is not only interesting in its own right but analytically notable that mid-eighteenth-century German authors of fables showed a particular tendency to choose songbirds as key protagonists to interact with humans. This is also plausible insofar as the potential of songbirds, especially nightingales, to

 | DOI:10.1163/9789004495395_008

inspire literature had already been recognised somewhat earlier, e.g. by Barthold Heinrich Brockes.[1]

This contribution highlights the theoretical debates that provided the basis for this poetic output and it further explains the special role of the songbirds in these Enlightenment fables, with special reference to human-animal competition. These observations also serve to clarify the general reasons why songbirds often feature in these tales. As will become apparent, the popularity of the songbird had many sources and implications, ranging from purely aesthetic considerations to moral, social and even political issues.

1 Animal Debates in Theories about Eighteenth-Century Fables

Whenever the roles of animals in German fables are discussed, Gotthold Ephraim Lessing's approach comes immediately to mind. Lessings *Abhandlungen über die Fabel* ('Treatises on the Fable', 1759) reached the well-known conclusion that animals in fables are denoted by their 'generally understood consistency of character' ("allgemein bekannte Bestandheit der Charaktere"). Since readers were deeply familiar with these reputed characteristics, they served to 'conjure up certain images inside us' ("ihre gewissen Bilder in uns erwecken") more easily than would complex individuals.[2] In other words, animals have a symbolic or, rather, iconic value within a fable: they 'stand' for something – e.g. certain human traits – that can effortlessly be conjured up without many words.

Despite its scholarly popularity, Lessing's approach was by no means representative of the functions which all animals were given by fable authors and theorists at the height of this genre's literary production. Rather than defining the 'typical' characters of certain animals, as later approaches in the vein of Lessing have done, the debates around 1730 and 1740 circled more around the relationships between humans and animals 'in general' and, based on that, tried to determine the value of the animal as a representative of human behaviour in each specific fable. The usage of animals in fables first and foremost created a problem for early Enlightenment poetical theory, because the fable provides its non-human protagonists with the uniquely human power of speech. This old genre characteristic of the fable directly challenged the concept of 'probability' (*Wahrscheinlichkeit*) as a central claim in Enlightenment aesthetic theory.

1 See Tanja van Hoorn: Brockes' Ornithopoetik. In: Mark-Georg Dehrmann and Friederike Felicitas Günther (ed.): Brockes-Lektüren. Ästhetik - Religion - Politik. Bern [et al.] 2020 (Publikationen zur Zeitschrift für Germanistik; NF Vol. 32), p. 167–183.

2 Gotthold Ephraim Lessing: Werke und Briefe in zwölf Bänden. Ed. by Wilfried Barner [et al.]. Frankfurt/M. 1985–2003. Vol. IV: Werke 1758–1759. Ed. by Gunther E. Grimm, p. 380.

The 'talking beast' in the fable thus needed to be defended in order to find its place in eighteenth-century literary production. This feature was one that the early theoretical works discussed most extensively. Johann Jacob Bodmer and Johann Jacob Breitinger, for instance, tried to solve the problem by defining fable-writing as an act of 'translation': Their idea was that God had left messages for humans in nature but only few people were able to understand them in their natural form. Accordingly, it was the poet's job to 'read' nature and to translate God's lessons into fables so that everyone had easy access to them. Talking animals in fables, therefore, only put into words what was already intrinsic to their characters but was difficult to decipher without speech.[3]

It was a widespread consensus in eighteenth-century literary theory that one of the main attractions of the fable sprang from the readers comparing animals to humans. The idea that animals might display more common sense and reason than did humans was first stated by Hieronymus Rorarius and passed on into the eighteenth century by Pierre Bayle[4] and John Locke. Several fables by Jean de La Fontaine portrayed animal behaviour in such a way that it could be seen as superior to human action. Inspired by La Fontaine, Friedrich von Hagedorn transferred this idea into the German fable and explicitly contrasted humans and animal in that way in fables like *Jupiter, die Thiere und der Mensch* ('Jupiter, the Animals and Man') or *Ulysses und seine Gefährten* ('Ulysses and his Companions'). Moral education based on this concept took place when readers saw creatures that are biologically 'beneath' them act in a more advanced and morally exemplary manner. Readers thus understand that they need to adjust their own conduct.

It is no surprise then that, according to Christian Fürchtegott Gellert (1715–1769), nothing gave more pleasure than the 'likeness the beast has with man' ("Aehnlichkeit, welche das Thier mit dem Menschen [...] hat").[5] This comparison, however, depended on two very important premises. Firstly, the animal's conduct must be as close to its natural behaviour outside the fable; it must be 'probable', especially since the power of speech was already stretching reality. 'Yet if we neglect the animals' nature and attribute to them too much

3 See Johann Jacob Bodmer and Johann Jacob Breitinger: Die Discourse der Mahlern. Vier Teile in einem Band. Hildesheim 1969 [Reprint of the ed. Zurich 1721–23], part III, XIX; Discours, p. 145–152.

4 See Hans Werner Ingensiep: Der Mensch im Spiegel der Tier- und Pflanzenseele. Zur Anthropomorphologie der Naturwahrnehmung im 18. Jahrhundert. In: Hans-Jürgen Schings (ed.): Der ganze Mensch. Anthropologie und Literatur im 18. Jahrhundert. DFG-Symposium 1992. Stuttgart and Weimar 1994 (Germanistische Symbosien Berichtsbände Vol. XV), p. 54–79, here p. 57.

5 Christian Fürchtegott Gellert: Gesammelte Schriften. Kritische, kommentierte Ausgabe. Ed. by Bernd Witte. Berlin and New York 1994 (Poetologische und moralische Abhandlungen, Autobiografisches Vol. 5), p. 19.

humanity, the likeness and along with it, the pleasure are lost' ("Sobald wir aber die Natur der Thiere verlassen, und ihnen zu viel Menschliches zuschreiben; sobald gehet die Aehnlichkeit und mit ihr das Vergnügen verlohren").[6] Secondly, the animals chosen by the author for these fables, must be those whose natural behaviour is fairly close to human behaviour. This requirement created a rather narrow selection; and it was one of the reasons why there was a decline, between 1730 and 1760/70, in the number of predominantly wild animals like wolves and foxes appearing in fables. Gellert even stated a clear preference for human protagonists to interact in fables, in cases where the risk of improbability became too great.[7] Consequently, Gellert's own fables made only limited use of the more traditional selection of animals and, ultimately, often ended up replacing them with human protagonists altogether.[8]

In the light of these concerns, it is apparent that, along with domesticated animals living close to humans,[9] the species which best met those criteria were songbirds. Songbirds have two advantages over most other types of animals: 1) they sing and are thus equipped with a kind of speech of their own; 2) their songs are often elaborate and, for this reason, can be considered as a form of art. This latter aspect means that songbirds can be viewed as artists and creators – giving them a key quality which makes comparison with humans extremely logical.

This perspective was all the more promising as people in the eighteenth century stressed the importance of musicality in the birds which they kept as pets, as Julia Breittruck has shown: The seventeenth century had already seen the emergence of attempts at writing down the scores of songbirds' tunes or even at teaching them new melodies.[10] Yet, Reimarus' famous 1760 study of

6 Ibid.

7 See Christian Fürchtegott Gellert: Gesammelte Schriften. Kritische, kommentierte Ausgabe. Ed. by Bernd Witte. Vol. 1: Fabeln und Erzählungen. Ed. by Ulrike Bardt and Bernd Witte in collaboration with Tanja Reinlein. Berlin and New York 2000, p. 21.

8 Handwerck elaborates on the declining number of animal fables in Gellert's three fables volumes in comparison to his early fables published in the perodical *Belustigungen* (see Hugo Handwerck: Studien über Gellerts Fabelstil. Marburg 1891, p. 7). – See also Peter Hasubek: Tugendspiegel und Lasterschelte. Gellerts Fabeln im soziokulturellen Kontext. In: Theo Elm and Peter Hasubek (ed.): Fabel und Parabel. Kulturgeschichtliche Prozesse im 18. Jahrhundert. Munich 1994, p. 157–174, here p. 165.

9 See for definitions and historical background Frederick E. Zeuner: Geschichte der Haustiere. Munich [et al.] 1967; Norbert Benecke: Der Mensch und seine Haustiere. Die Geschichte einer jahrtausendealten Beziehung. Stuttgart 1994.

10 See Julia Breittruck: Musikalisierte Tiere. Die Entdeckung der Vogelstimmen in der Frühen Neuzeit. In: Marianne Sommer and Denise Reimann (ed.): Zwitschern, Bellen, Röhren. Tierlaute in der Wissens-, Medientechnik- und Musikgeschichte. Berlin 2018, p. 21–38.

animal instincts, which had a particular and innovative focus on their 'aesthetic drives', did not elaborate on songbirds and their acoustic production.[11]

2 Songbirds and Humans

Birds were by no means new to the fable during the eighteenth century. The traditional inventory included, for instance, the king of the frogs (the stork), the crow with peacock feathers, or the raven who fights the fox over a piece of cheese.[12] However, birds which did nothing but sing were considerably rare, with the exception only of the nightingale.[13] This situation, however, changed significantly in mid-eighteenth-century German fables, as their titles indicated. The nightingale in particular became universally common and was often paired with some other bird species. The titles include: Gellert's *Die Nachtigall und die Lerche* ('The Nightingale and the Lark'), *Die Lerche und die Nachtigall* ('The Lark and the Nightingale'), *Der Kukuk* ('The Cuckoo') and *Die Nachtigall und der Kukuk* ('The Nightingale and the Cuckoo'), Johann Georg Bock's *Der Guckguck und die Nachtigal* ('The Cuckoo and the Nightingale'), *Die überdrüßige Nachtigal* ('The Weary Nightingale'), *Die Wachtel und die Amsel* ('The Quail and the Nightingale'); Johann Adolf Schlegel's *Die eitle Nachtigall* ('The Vain Nightingale'), *Die Eule und die Nachtigall* ('The Owl and the Nightingale'); and Magnus Gottfried Lichtwer's *Der Uhu und die Lerche* ('The Eagle Owl and the Lark').[14]

In order to illustrate the nature of these fables in greater depth, the analysis now turns to take a closer look at some examples. The first fable to probe is Ludwig Gleim's *Die dankbare Nachtigall* ('Grateful Nightingale') from 1758:

---

11 See the list in Hermann Samuel Reimarus: Allgemeine Betrachtungen über die Triebe der Thiere, hauptsächlich über ihre Kunst-Triebe. Zum Erkenntnis des Zusammenhanges der Welt, des Schöpfers und unser selbst. Hamburg 1760, p. 143–150.

12 See for an overview Gerd Dicke and Klaus Grubmüller: Die Fabeln des Mittelalters und der frühen Neuzeit. Ein Katalog der deutschen Versionen und ihrer lateinischen Entsprechungen. Munich 1987.

13 Cf. Hesiod: Theogonie. Werke und Tage. Griechisch und deutsch. Ed. by Albert von Schirnding. Zurich 1991, p. 201–212.

14 Johann Georg Bock: Der deutsche Aesop. Bestehend in dreyhundertvierundzwanzig lehrreichen Fabeln, welche in gebundener Schreibart entworfen und als moralische Wochenblätter stückweis ausgeliefert worden. Königsberg in Preußen 1743; Johann Adolf Schlegel: Fabeln und Erzählungen. Faksimiledruck nach der Ausgabe von 1769. Mit einem Nachwort von Alfred Anger. Stuttgart 1965; Magnus Gottfried Lichtwer: Vier Bücher Aesovischer [!] Fabeln in gebundener Schreib-Art. Leipzig 1748.

Die dankbare Nachtigall

Ein Falke sah, mit großen wilden Augen
So hell, als wie Kristall,
Nach einer liebe Nachtigall,
Und drohete, das Blut ihr auszusaugen!

Zwar sah sein Aug', auf Lande neben Hecken
Auch einen jungen Staar,
Allein sein Mord=Gedanke war:
Die besser singt, die muß auch besser schmecken.

Er wollt', allein, er musste sichs begeben,
Der Mörder! denn ich schoß,
Mein Schießgewehr schnell auf ihn loß,
Und traf ihn recht, und rettete ihr Leben.

Nun hüpft, (komm Damon, es zu sehen!)
Die kleine Sängerin,
Wenn ich in meinem Garten bin,
Um mich herum, und singt in der Alleen![15]

The Grateful Nightingale

A hawk looked out, with huge wild eyes
As bright as a crystal,
For a dear Nightingale
And threatened to drink her blood!

In fact his eye too saw, hidden in hedges,
A young starling.
Yet the murderer thought:
The one that sings better must taste better.

So rapidly flew – like a thunderbolt,
Straight as an arrow,
Just as her trill was heard –
The hawk to fetch his dinner.

---

15 Johann Wilhelm Ludwig Gleim: Fabeln. Berlin 1758, p. 31–32. – Translations are mine unless otherwise noted.

He aimed, yet was forced to yield,
The butcher! for I fired
My shotgun at him in time,
And hit *him*, and saved *her* life.

Now flutters (come, Damon, to see!)
The little singer
Around me when I am in my garden
And fills the avenue with song!

Gleim's fable recalled the tale of hawk and nightingale as told by Hesiod.[16] In this classical fable, the nightingale begs for her life and refers to the beauty of her song as an argument that her life should be spared. Yet the hawk remains unconvinced and eats her anyway. Gleim recounted a variation of that story as the action is interrupted by the speaker (a hunter?) who saves the nightingale by killing the hawk. This adaptation changed the fable's message significantly. While Hesiod's version made it clear that the production of art cannot save the nightingale's life, Gleim allowed sophistication to triumph over the archaic 'eat or be eaten'; and the new outcome happens though human intervention. The human thus appears as the nightingale's ally and, in return, gets to enjoy the bird's singing in the garden – a result that would be without meaning for the more 'primitive' hawk. In this way, humans and songbirds are united in refinement, represented respectively by the production and the appreciation of art.

Moreover, the nightingale seems all the more human as the title attributes to her the emotion of gratitude. The fact that the bird's saviour is rewarded for a good deed by beautiful music is not a result of mere chance but, as indicated by the fable's title, must be read as an act of compensation which the nightingale deliberately choses to make. Gratitude and reimbursement are signs of the nightingale's high level of civilization and thus her likeness in character to that of humans. The songbird is a social being, a 'dear nightingale', whereas the hawk appears as a selfish 'murderer'. In this manner, Gleim's 'Grateful Nightingale' confronted the traditional image of the wild animal (the hawk) with that cultivated human-like songbird, represented by the nightingale. Thus he indicated a clear preference for a cultivated and, so to speak, humanized nature.

16 See for notes on German fables drawing on Hesiod Bernd Witte: Emblematische Bilder. Die deutschsprachige des achtzehnten Jahrhunderts zwischen Oralität und Literalität. In: Daphnis 25 (1996), no. 4, p. 715–738, here p. 730ff.

3 Songbird Interaction

In order to determine the specific function of the songbird in eighteenth-century German fables, it is also fruitful to examine the interaction of several songbirds with each other, as such encounters are an essential element of many songbird fables. Looking at this theme, a pattern emerges which is characteristic of many fables from the period, with only slight individual variations. A good example is the 1744 fable by the Swiss poet Ludwig Meyer von Knonau. entitled *Die junge Schwalbe* ('The Young Swallow'):

Die junge Schwalbe

Den Ton aus der so kleinen Brust
Der liederreichen Nachtigallen
Hatt' eine junge Schwalbe Lust
Nach ihrer Mundart nachzulallen.
Ey! Mutter, könnt ich auch so singen,
Sprach sie, wie süsse würd es klingen!
Wie hörte man den Klang so weit?

Die Alte sprach: Kind, schick dich in die Zeit;
Bleib eine Schwalbe, wie du must,
Und hemme deine bange Lust;
Hätte es dem Jupiter gefallen,
Wir wären alle Nachtigallen.
Man soll der Lüste Trieb bey Kindern
So viel es möglich ist, verhindern.
Mit seinem Stand zufrieden leben,
Ist wohl das nützlichste Bestreben.
Betrachte so die viele Stuffen,
Die Amsel, wozu sie beruffen,
Wozu der buntgefärbte Specht,
Wozu der Sperlinge Geschlecht,
Wie andre nur nicht schwatzen können,
Und auch dein Zwitschern dir mißgönnen.

Gut, sprach das Schwälbchen, ich verehre
Mit Ehrerbietung deine Lehre,
Die du mir unerfahrnen giebst,
Du thust es, weilen du mich liebst.

Ach! wüßt ich sie recht auszuüben;
Dabey dich als ein Kind zu lieben![17]

The Young Swallow

The chant from the nightingale's tiny bosom,
A bird so rich in melodies,
Made a swallow aim to try
And produce an echo in her tongue.
Ay! Mother, if only I could sing thus,
She said, what sweet tunes I would create!
And all through the land I would be heard!

The elder said: Child, restrict yourself and be content;
Remain a swallow as you must,
And tame your fretful yearning;
Had Jupiter pleased to make but nightingales,
We would all be nightingales.
It's wise to curb our children's lust
Within the frame of possibilities.
To be content with one's own rank
Is the most beneficial undertaking.
Consider all the steps and stages,
The blackbird in his calling,
The great spotted woodpecker in his,
The line of sparrows in theirs.
Where others aren't allowed to chatter,
They will begrudge your chirrup.

'Tis well, the little swallow said, I honour
Your lecture in deference,
With which you have rewarded the naïve.
You teach me out of love;
Alas, could I apply it well
And, at the same time, love you as a child should!

17 Ludwig Meyer von Knonau: Ein halbes Hundert neue Fabeln. Zurich 1744, p. 3–4.

It is clear that this fable must be read as a commentary on the question of social rank. The mother swallow conveys very common advice for the period, as she tells her daughter to stick with her own class instead of aiming for social advancement that would only create envy. This explanation, however, only captures the surface of what is implied in Meyer von Knonau's fable. It is again worth comparing this plot with the traditional role of animals in fables. The well-known animal kingdom, with the lion at its top, is replaced by a society of (song-)birds, whose most superior member is the nightingale. Yet, the nightingale's outstanding position does not stem from her surplus of physical strength, which gives her power over the other members of the community. Her superiority is instead founded upon her excellent singing, her aesthetic output. This imagery mirrored the paradigm shift that can be detected in eighteenth-century political theory, which increasingly aimed at a leadership based on intellectual or moral superiority: merit instead of heritage. The nightingale's prime position is cause for envy in many eighteenth-century fables, which nevertheless seek to justify the bird's elevation. Meyer von Knonau's book of fables, whose third 1757 edition included illustrations, not only exemplified this new perspective in its traditionally programmatic opening fable *Die junge Schwalbe* ('The Young Swallow', figure 7.1) but also gave a contrasting view in *Die Vögel und die Nachtigall* ('The Birds and the Nightingale', figure 7.2).

This latter fable on the surface appears to be a testament to equality, as the nightingale defends herself against the other birds' criticisms by highlighting the value of each bird's individual voice. Yet the fable's key message is a traditional one, that there is no reason to question the existing hierarchy established by 'Jupiter's law' ("Jupiters Gesetze"):

> Vergeßt nur euern schwarzen Neid,
> So fällt denn aller Unterscheid
> Gleich zwischen euch und euern Brüdern,
> Und zwischen unser aller Liedern.[18]

> Put down your black envy,
> Thus all inequality subsides
> At once between you and your brothers
> And between all our songs.

18 Ibid., p. 18.

F. 1.

FIGURE 7.1 *Die junge Schwalbe* ('The Young Swallow', 1757). In: Ludwig Meyer von Knonau: Neue Fabeln. Dritte und verbesserte Auflage. Mit LVIII. Kupferstüken. Zürich 1757, Fig. 1 [unpag.].

Given this specific context, it is no surprise that the focus of most eighteenth-century fables highlighted the theme of two or more songbirds in competition

FIGURE 7.2 *Die Vögel und die Nachtigall* ('The Birds and the Nightingale', 1757). In: Ludwig Meyer von Knonau: Neue Fabeln. Dritte und verbesserte Auflage. Mit LVIII. Kupferstüken. Zürich 1757, Fig. VIII [unpag.].

with each other. This motif is, for instance, also the case in Gellerts *Die zwei Schwalben* ('The Two Swallows'), in which the birds do their best to outshine each other and ask the lark to be the judge of who sings more beautifully. Of course, the lark's verdict can only be that the two swallows are on the same level and will never do better than the nightingale.[19]

On the one hand, fables such as this one displayed the ideal of conviviality[20] (*Geselligkeit*) and offered a critical discourse on the quality of aesthetic production. Eighteenth-century German fables were full of self-declared 'experts' and authorities mirroring the contemporary surge of literary criticism.[21] In exposing ill judgement, the fables attempted to alert the reader to his or her own prejudices and thus made it their goal to teach self-criticism and objective judgement to a broader audience. The discussion of matters of moral and aesthetic judgment refers to the concept of *moral sense*[22] and it usually takes place in the context of self-love because that emotion can lead to self-deception and so drastically impair the impartiality of one's judgment.[23]

On the other hand, while the mother swallow in Meyer von Knonau's text instructs her daughter to acknowledge the superiority of the nightingale, the wise mother also advises her child to appreciate the diversity of tunes in the world of songbirds, in which every bird family has its own unique voice to contribute. The use of the term 'dialect', which the Swiss author chose in possible reference to the linguistic varieties of his home country, expressed that twin notion of equality *and* diversity very clearly. Enlightenment literature generally stressed the value of what is 'natural' to a certain gender, age, rank or regional descent. No one should be forced or force himself or herself to do something that is not part of his or her nature.

Hence that recommendation was the other key lesson which the songbird fables illustrated time and time again. One of Gellert's early fables told the story of a lark striving to sing as well as the canary. This time it is the nightingale's

19 Gellert, Gesammelte Schriften, Vol. I, p. 134.

20 See Emanuel Peter: Geselligkeiten. Literatur, Gruppenbildung und kultureller Wandel im 18. Jahrhundert. Tübingen 1999 (Studien zur deutschen Literatur Vol. 153).

21 See Jürgen Habermas: Strukturwandel der Öffentlichkeit. Untersuchungen zu einer Kategorie der bürgerlichen Gesellschaft. 1st ed., Frankfurt/M. 1990 [Reprint of the ed. Neuwied 1962], p. 86–121, esp. p. 103, note 32.

22 See on this and on other backgrounds for eighteenth-century German fables in the context of moral education through literature Kristin Eichhorn: Die Kunst des moralischen Dichtens. Positionen der aufklärerischen Fabelpoetik im 18. Jahrhundert. Würzburg 2013.

23 See Friedrich Vollhardt: Selbstliebe und Geselligkeit. Untersuchungen zum Verhältnis von naturrechtlichem Denken und moraldidaktischer Literatur im 17. und 18. Jahrhundert. Tübingen 2001 (Communicatio Bd. 36), p. 3.

turn to inform the lark of her foolishness: “Stick by your tune, your natural dialect;/ In your tune you will shine, in his fail terribly”.[24] Everyone needs to develop his or her particular talent. To fall silent in view of the nightingale's superiority would be just as silly as trying to imitate her, as is the siskin's lesson in Gottlob Wilhelm Burmanns *Die Finken und der Zeisig* (‘The Finches and the Siskin’; 1773):

> O rief ihr Mentor, welche Grillen!
> Ihr schweigt also, weil ihr nicht Nachtigallen seyd?
> Und schweigt aus Eyfersucht und Neid?
> Lernt besser eure Pflicht erfüllen;
> Singt wie die Finke singen muß;
> Denn Finken können nicht die Gaben
> Berühmter Nachtigallen haben –
> Das ist des Himmels weiser Schluß.[25]
>
> O, their mentor called, what idiocy!
> You fall silent since you are not nightingales?
> You fall silent out of jealousy and envy?
> Understand your obligation:
> Sing as the finches must sing;
> For finches are not blessed
> Like famous nightingales –
> It is the Lord's decision in his wisdom.

The final two lines of this fable put its message in a nutshell: “Alas, defeat your jealousy/ And learn to sing beautifully in a lower category!” (‘*O lernt die Eyfersucht bezwingen:/ Und schön in niedrer Spähre* [!] *singen!*’).[26]

Consequently, confident songbirds defy the call to imitate another's voice, as the lark does in Bock's *Der Deutsche Aesop* (‘The German Esopus’; 1743) as well as in Lessing's *Die Nachtigall und die Lerche* (‘The Nightingale and the Lark’), to name a few more examples.[27] Burmann's fable, quoted above, also portrayed such behaviour as an act of heresy.

The discursive field of the tables examined here nevertheless offers a justification of the current situation. Meritocracy, which all the fables emphasize in

---

24 Gellert, Gesammelte Schriften, Vol. I, p. 14.

25 Gottlob Wilhelm Burmann: Fabeln und Erzählungen in vier Büchern. Berlin 1773, p. 79–80.

26 Ibid., p. 80.

27 Lessing, Werke und Briefe IV, p. 329 (Fab. III, 2).

unison, makes it hard to argue in favour of a rigid class society. It raises ambitions in those who find themselves at the bottom of the social hierarchy. This theme, after all, is the phenomenon portrayed in fables like *Die junge Schwalbe*, *Die zwei Schwalben*, or *Die Finken und der Zeisig*. However, the important thing to note here is that the aspirations expressed by those songbirds which are not nightingales are always, and without question, discouraged. On the one hand, this outcome does lead to a validation of the individual voice – a bird may not be able to sing as beautifully as the nightingale, but it could, and should, still participate in the communal song. On the other hand, this validation of *diversity* confines the songbird in question to his or her particular place on the social ladder somewhere *beneath* the nightingale. In other words, the songbird state does nothing to create *equality* but upholds yet another God-given, hence unchangeable, *hierarchy*.

On closer inspection, therefore, the ideal of meritocracy is actually quite hollow. For birds cannot become better singers by constant practice or rise to the top by displaying talents other than singing. Just as in the old animal kingdom, where physical strength was the main criterion for class distinction, birth determines forever the individual's place in society – and it is seen as foolish to strive for another place. This analysis thus reveals that the songbird fables obviously used the increasingly popular ideal of meritocracy as a front to defend the social status quo and to create what might be called a state of 'pseudo-equality'.

This conclusion is all the more visible in fables such as Johann Adolf Schlegel's *Die Krähe und die Nachtigall* ('The Crow and the Nightingale'), which do not even tell of competition but merely state the obvious superiority of the nightingale – all the more convincingly as the nightingale's challenger is not even a songbird but a jealous critic with no legitimacy. Hence the nightingale declares: "You may call me a bungler:/ Yet, your judgement does not impress me;/ For those who cannot praise cannot condemn" ('Und solltest du mich auch den größten Stümper nennen:/ So rührt mich doch dein Urtheil nicht;/ Denn, wer nicht loben kann, wie soll der schimpfen können').[28] Apart from the sideswipe at literary criticisms in this fable, which would be worth being investigated in a separate study,[29] Schlegel presented the reader with yet another story of a bird's overestimation of its own abilities. The crow, just like the young swallow in Meyer von Knonau's text, needs to be reminded to acknowledge the

28 Johann Adolf Schlegel: Fabeln und Erzählungen. Leipzig 1769, p. 150.

29 See Helmuth Kiesel and Paul Münch: Gesellschaft und Literatur im 18. Jahrhundert. Voraussetzungen und Entstehung des literarischen Markts in Deutschland. Munich 1977.

nightingale's dominance. So the fable provided one more negative example of a living being not knowing its limitations and, ultimately, its place.

4 Conclusion

What conclusions can be drawn from this close examination of the roles of songbirds in mid-eighteenth-century German fables? As has been shown, songbird fables reflected many big contemporary themes: including changes in aesthetic theory, wider philosophical debates from moral philosophy to the philosophy of state and society. and – not least – developments within natural philosophy in accordance with new perceptions of animals in the works of Pierre Bayle, John Locke, and Hermann Samuel Reimarus.

The first factor that worked in favour of the songbird as a fable protagonist was an aesthetic idea: the concept of probability. Since animals in fables were intended to represent human behaviour, fable authors and theorists started to look for animals which 'resemble' humans. Hence the songbird's elaborate musical outpouring could be understood as a logical parallel of human speech. Furthermore, a second point can be made: the songbird was seen as an artist, its song an aesthetic production, in parallel with human speech. Hence songbirds were not only taken as resembling humans generically, they were also taken as the animal representatives of *sophisticated* humans. The songbird state portrayed in the fables embodied the highly civilized society which mid-century Enlightenment theorists usually claimed for their own time in comparison with earlier and more barbaric eras.[30] Thirdly, too, the songbird served as a proxy for *modern* man in particular; and the songbird state, replacing the old animal kingdom of traditional fables, offered an allegory of a particularly *modern* society.

Given this shift, it was not surprising that most Enlightenment ideals were in some way or other present in fables written during the second third of the century. Ideas such as conviviality, equality, diversity and meritocracy play an especially huge role as the birds debate matters of status and priority. However, the songbird fables contained not only a modern streak in the sense that they incorporated recent concepts and debates in order to 'update' the old genre of the fable. They also displayed a much more conservative, yet rather typical,

30 This portrayal is a bit simplified, of course, in order to emphasize the paradigm shift in the second half of the century which already influences the different approach to animal fables in Lessing's theory and fable production.

element of the German Enlightenment, since the ideas of meritocracy and diversity did not include the chance to climb up the social ladder.

On the contrary, the social hierarchy remains just as fixed as before; and the songbird fables make it an explicit point to warn their readers constantly against any foolish notions that could call the status quo into question. In this way, these fables very much constituted a response to the perceived danger that some people – just like Meyer von Knonau's young swallow or the other birds in fruitless competition with the nightingale – might argue for a more radical change of system given that some changes were already happening. These voices are not silenced in the fables but are clearly shown their own limitations. Ultimately, therefore, birds – and humans – must learn to content themselves with the social position into which they are born.

Finally, this conclusion does not mean that eighteenth-century fables simply presented an 'outdated' ideology as older Marxist and sociohistorical studies have claimed.[31] Instead, the fables need to be interpreted within the context of the unfolding debates of the time. Continuing research will no doubt reveal further intricacies. Yet the significant contribution of the songbirds to eighteenth-century aesthetic debates and to contemporary observations on human-animal relations is apparent: their capacity for song highlights the tensions between competition and conservatism within a changing society.

31 See Elisabeth Herbrand: Die Entwicklung der Fabel im 18. Jahrhundert. Versuch einer historisch-materialistischen Analyse der Gattung im bürgerlichen Emanzipationsprozeß. Wiesbaden 1975.

CHAPTER 8

The Talking Parrot: Brazilian National Symbol and Avatar of Human Identity for John Locke

Antônio Carlos dos Santos

Résumé

Le but de cette contribution est d'analyser la représentation du perroquet au Brésil entre les XVIe et XVIIe siècles, en se basant sur les relations de voyage françaises et néerlandaises en terres brésiliennes et comment elles ont aussi marqué l'imaginaire culturel brésilien. Ces récits ont popularisé l'animal non seulement comme exotique, mais aussi comme celui qui ressemblerait le plus à l'âme du peuple brésilien. Cet article est divisé en deux parties: dans la première, nous évoquons les marques historiques de la présence du perroquet au Brésil, dans la seconde, nous nous consacrons à l'analyse philosophique de cet animal dans l'optique proposée par John Locke (1632–1704). Ce dernier fait écho à des préoccupations actuelles sur cet animal.

Abstract

This contribution analyses representations of the Brazilian parrot between the sixteenth and seventeenth centuries. It is based upon accounts, written by French and Dutch travellers in Brazil, who strove to understand Brazilian popular culture and its representations. Their reports popularised the parrot not only as an exotic bird but also as the one which they felt most resembled the soul of the Brazilian people. The ensuing discussion is divided into two sections: the first considers historical evidence for the presence of the parrot in Brazil; the second examines the philosophical analysis of this remarkable bird, with its famed powers of mimicry, from the perspective proposed by the English political theorist John Locke (1632–1704). The parrot thus constitutes a reminder of continuing philosophical debates about species identity as well as a lively Brazilian cultural symbol.

1 Introduction

Culturally, there are many attempts at representing the spirit of a people through an animal. Examples include the eagle as the symbol of the USA;[1]

1 The eagle was chosen as a symbol of the US in the eighteenth century. There is an amusing letter by Benjamin Franklin to his daughter Sarah where the founder of America indicated that

 | DOI:10.1163/9789004495395_009

the rooster, as the archetypal representative of France; the condor, of Chile; the rooster of Barcelos, in Portugal; the kangaroo, of Australia, among others. In that context, the parrot is certainly the animalistic translation of Brazilian national identity. Unlike these other cases, however, this relationship does not occur because of positive aspects of the animal's own virtues but, above all, because it has disreputable characteristics. In Brazil, the parrot is understood as a shifty, playful, rascally creature, among other ignoble traits. Maybe that is why this national bird does not appear in any official documents, but instead it is omni-present in popular culture. How, historically, did the parrot come to be associated with the image of Brazil? Why is this animal not admired for its virtues, but for its vices? And what is the relationship between this iconic role and philosophical debates about the distinction between humans and animals? This contribution addresses these issues, which appear separate but have a common root in responses to the talking parrot.

The first section of the analysis explores the representation of the parrot in Brazil between the sixteenth and seventeenth centuries, taking two references that became embodied in the Brazilian cultural imaginary. They were French and Dutch reports of experiences in the Brazilian territory during this period – reports which popularized the animal not only as exotic, but also as the one that seemed best to represent the soul of the Brazilian people. Then, having explored the historical references to the parrot's presence in Brazil, the second section will investigate the philosophical analysis of this animal according to the perspectives advanced by the English philosopher John Locke (1632–1704).[2] The issues raised thereby are ones that resonate to the present day.

2 Historical Evidence of Parrots in Brazil

Brazil, perhaps, is one of the few countries in the world that has a 'birth certificate'. It took the form of a letter, written on May 1, 1500 by Pero Vaz de Caminha, and destined for the king of Portugal, Dom Manuel I. Caminha was the clerk

---

the turkey embodies more of the characteristics of the American nation than the Eagle: "For my own part I wish the Bald Eagle had not been chosen as the Representative of our Country. He is a Bird of bad moral Character. [...] For in Truth the Turkey is in Comparison a much more respectable Bird, and withal a true original Native of America". – The letter is unpublished in print; online: https://franklinpapers.org [14.06.2021].

2 According to the conventions adopted by the Locke Society, quotations from his *Essay on Human Understanding* [henceforth Essay] follow this pattern: the Roman numerals indicate the book, the first Arabic numerals, the chapter, and the second, the paragraph.

of the fleet commander Pedro Álvares Cabral, who led the thirteen ships that introduced Brazil to the Europeans. Among the ship's company, Roberto P. Toledo recorded that, since the arrival of the ship with the first Portuguese, the parrot was ever present among them.[3] In this letter, rich in details and information about the exact moment when the land was first sighted by the Portuguese, there were five mentions of parrots. One referred to the impressive beauty of one type of parrot which was briefly described. At the same time the author noted that one of his colleagues wanted to acquire the bird from the indigenous people in return for some trinket brought by the Portuguese. Indeed, according to Toledo, in 1501 the first ship back to Lisbon, named 'The Annunciation', carried numerous Brazilian parrots, which were greeted with great fascination in Europe. Colourful, exuberant, different from European animals, the parrot had another characteristic that was remarkable, besides its beauty: it imitated the human voice.

As shown by Alberto Cantino's world map of 1502 – the first in which Brazil appeared – there was a trio of colourful parrots that decorated the newly annexed territory. Influenced by this map, the new Portuguese lands were identified as '*Terra Papagalli*'. In this representation, the exuberance of the colourful animal was associated with its ability in vocal expression, which stirred great curiosity and encouraged the search for this type of animal throughout Europe.[4] Hence the founding document of the Brazilian national territory served to disseminate to the world an image of a hitherto unknown paradisiacal place, where natural wealth was abundant and which (it seemed) anyone could appropriate. Such ideas soon led others, in addition to the Portuguese, to take an interest: notably the French and the Dutch.[5]

Much of this ambition was displayed in travel narratives of the period, especially in those by Jean de Léry, André Thévet,[6] Yves

3 Roberto P. Toledo: Papagaio! A tradução ornitológica da nacionalidade. In: Piauí 1 (2006); online: https://piaui.folha.uol.com.br/materia/papagaio [03.06.2021].

4 Several reports, besides the Caminha Letter, comment on animal life in the new ries; see, e.g., Fernão Cardim: *Treatises of the Land and People of Brazil* (written between the end of the sixteenth and the beginning of the seventeenth century); Pero de Magalhães Gandavo: *History of the Province of Santa Cruz* (1576); and Friar Vicente do Salvador: *History of Brazil* (1627).

5 Frank Lestringant: De Jean de Léry a Claude Lévi-Strauss. Por uma arqueologia de Tristes trópicos. In: Revista de Antropologia 43 (2000), no. 2, p. 81–103.

6 André Thévet was a cartographic advisor to the King of France, Henry II. He went to Rio de Janeiro, in Admiral Nicolas Durand de Villegagnon's fleet, remaining in Brazilian lands from November, 1555, to January, 1556, observing the nature and indigenous people of Guanabara Bay. He was the chief populariser of the expression 'Antartic France', referring to the French colonial experience, when he publishied *Les singularitez de la*

d'Evreux,[7] Claude d'Abbeville,[8] and Hans Staden,[9] which remain the most remarkable and popular in Brazil. This vast literature was written by wanderers, missionaries, or simply adventurers who resided among the Brazilians and who brought their contribution to the study of their origins from different perspectives. Here the focus falls upon two historical literatures: those of the French and the Dutch.

From France, as is well known, the first narratives of travel to the New World were recounted in the book *Viagem à terra do Brasil*, published in 1578 by Jean de Léry.[10] It offered in effect a sort of founding myth, especially relating to the settlement led by Villegagnon, in what became known as the 'project of Antarctic France'. The book contained precious ethnographic information, as well as pre-anthropological reflections, which was so detailed that Claude Lévi-Strauss described the volume as the "ethnologist's breviary".[11] Jean de Léry, as is well known, was part of the group of fourteen Calvinists who left Geneva to participate in the attempt at establishing a French colony in Guanabara Bay, just outside Rio. Despite having promised religious freedom to Protestants (the colony was composed of Catholics and Protestants, made up of Europeans from various nationalities), the leader Nicholas Durand de Villegagnon quickly began to challenge their beliefs and proceeded to persecute them. After eight months, the Protestants decided to leave the colony and spent two months living on the mainland, near the indigenous Tupinambá people of that region. It was during this period in Brazil, between February 1557 and January 1558, that Léry proved to be a remarkable observer of local nature and indigenous customs.

France Antarctique (Paris 1557), illustrated with 41 woodcuts. There is a Brazilian translation: André Thévet: As singularidades da França Antártica. São Paulo 1978.

7 Yves d'Évreux was a French cleric, who participated in the expedition sent in 1612 to Maranhão, with his friend Claude d'Abbeville. Returning to France, he published in 1615 *Voyage dans le nord du Brésil.*

8 Claude d'Abbeville was a French Franciscan friar, who participated in the expedition sent in 1612 to Brazil (Maranhão) by the government of his country. He is the author of *Historie de la mission des pères capucins en l'isle de Maragnan et terres circonvoisines* (1614).

9 Hans Staden was a German mercenary adventurer who was in Brazil twice. The last time, he spent nine months as a slave of Tupinambá people. Back in Germany, Staden wrote in 1557 a report about his travels to Brazil (*Warhaftige Historia und beschreibung eyner Landtschafft der Wilden*) that became a great success at the time. – Cf. Hans Staden: Duas viagens ao Brasil. São Paulo 1974.

10 Jean de Léry: Histoire d'un voyage faict en la terre du Brésil [1578]. 2ième édition, 1580. Paris 1994. The version used in this article was the Brazilian one: Jean de Léry: Viagem à terra do Brasil. Tradução e notas de Sérgio Milliet. Rio de Janeiro 1961.

11 Claude Lévi-Strauss: Tristes trópicos [Fr. *Tristes tropiques*, 1955]. São Paulo 2009, p. 87.

After detailing the life of the Tupinambá people, in one of the remarkable moments of the book, Léry describes the types of parrots; he was impressed with their variety, and records:

> they are notable for the beauty of the plumage and how, when taught, they are the ones who speak best, they appreciate us more than others. In fact, an interpreter presented me with one of these birds that three years ago He kept in his possession. She pronounced so perfectly the words of the wild language and the French language that it was not possible to distinguish her voice from that of a man.[12]

Léry's description, as cited above, was impressive because he made no distinction between the language of the 'savage' and that of Michel de Montaigne: he simply admired the perfection of the bird's speech, combined with its beauty. Following this dialogue, Léry recorded his encounter with an indigenous woman, who had a parrot, and used it as her bargaining counter. He stated:

> Greatest wonder still seemed to me, however, a parrot of this species belonging to a certain Indian from a village distant two leagues from our island. It would be said that this bird understood what its owner said to him. When we passed by this one asked us: "give me a comb or a mirror and I will make my parrot sing and dance in your presence". If we gave what was asked, all it took was one word for the bird to start jumping on the stick it landed on, talking, whistling and teasing the savages from the start to war [!], in an incredible way. And when the owner said to sing, he sang; and also danced when she commanded him. If, however, we did not give him anything, she merely said asperously to the parrot: *augé,* that is, "stop" and he would [...] [behave accordingly] without uttering a word and [...] [no matter what we said,] he did not move either his foot or his tongue. If the ancient Romans were wise enough to make sumptuous funerals to the crow who, in their palaces, greeted them by their own names, even taking the life of those who killed him, as Pliny [...] [relates], imagine what they would not have done if they had possessed a parrot so perfectly taught![13]

Still following this conversation, the Calvinist wanted the parrot for himself, in exchange for any object that the indigenous woman requested: yet she did

12 Léry, Viagem à terra do Brasil, p. 123.

13 Ibid.

not respond or ask for a big gift, which, given Léry's limited possessions, would have been simply impossible. The dialogue ended abruptly, but the reader is left to understand the problem: any mercantile conversation involving that precious parrot was out of the question. For the indigenous woman, there were priceless things, and that animal was one of them.

It was also true that the entire work referred to various types of parrots and the way in which the indigenous people related to them and to the rest of nature. When deciding to return to France, Léry resolved, like his other shipboard companions, to take with him several parrots of different species. But at the conclusion of the work, he reported that the journey was marked by much suffering. The boat was already damaged when it left Rio de Janeiro; the food supply was too small in relation to the amount of people on board; and the length of the trip extended much longer than expected. When therefore he was already near the end of the route, with no food available, Léry recorded, with great sadness, that he was forced to consume the very parrot that he would have taken as a gift for an admiral. This episode happened not only with respect to the Calvinist Léry, but also to several of his shipboard companions who also carried parrots to traffic in Europe. The high financial value of such an animal was worth such sacrifice only in the extreme circumstances of sheer survival.

Through such detailed records involving the parrot, the reader gains a better understanding of the situation. It was clear that the parrot had become an exotic product of a highly profitable trade, because it was an exuberant animal, funny, maliciously smart, and valued for its ability to imitate the human voice. Its rarity in Europe made it expensive, so that presenting a parrot as a gift was like offering precious stones.

Undoubtedly, however, the most striking episode involving the parrot occurred during the Dutch occupation of north-eastern Brazil. It all started in 1630, when an ambitious group of the *Geoctroyeerde West-Indische Compagnie* (WIC) invaded Recife, a region considered to be the richest sugar producer in the world at the time, and settled there for 24 years. The area became known as 'Dutch Brazil', with John Maurice, Prince of Nassau-Siegen, named by the Dutch as governor-general. One of the most impressive aspects of the history of this German prince in north-eastern Brazil was his concern to update and to 'civilize' what seemed to him a very primitive place.[14] Contrary to the

14 Maurice of Nassau brought several scientists to Brazil, among them Georg Marcgraf and Willem Piso. The first, born in Liebstadt, Germany, in 1610, studied the climate and classified hundreds of species of Brazilian fauna and flora. He was a polymathic naturalist, astronomer, mathematician, surveyor, and cartographer. The second one was born in the Dutch town of Leiden in 1611. He was a naturalist and Nassau's physician,

traditional monoculture, he forced the local farmers to plant manioc which, as he said, would become the 'bread of Brazil'. He also forbade the farmers to fell the cashew trees, in order to preserve them from extinction. In addition, Nassau built palaces, bridges, and roads. An encourager of the arts, he promoted plays as well as many artistic and cultural manifestations. Tolerant, he allowed the construction of synagogues, on the one hand, and promoted a policy of rapprochement between Calvinists and Jews, on the other.[15]

It was within this historical context that an episode involving a parrot and the Dutch governor in Brazil was recorded in the Memoirs of William Temple, published in 1689,[16] which referred to a dialogue maintained between him and Maurice of Nassau. The encounter was widely circulated, to such an extent that, on the threshold of the eighteenth century, the English philosopher John Locke recapitulated all the original dialogue in book II, chapter 27, of the fourth edition (1700) of his *Essay Concerning Human Understanding.* To prove that this event had actually occurred, in addition to the text by William Temple, there is also a surviving painting of such a parrot, plus a long sentence, next to Nassau's signature (a testament which will be further examined in the second section of this discussion).

With reference to the encounter itself, John Locke's report drew upon Temple's original text:

> I [sc. Temple] had a mind to know from *Prince Maurice's* own Mouth, the account of a common, but much credited Story, that I had heard so often from many others, of an old *Parrot* he had in *Brasil*, during his government there, that spoke, and asked, and answered common Questions like a reasonable Creature [...] I had heard many particulars of this Story, and assevered by People hard to be discredited, which made me ask *Prince Maurice* what there was of it. [...] when he came to Brasil, and though he believed nothing of it, and it was a good way off, yet he had so much Curiosity as to send for it, that it was a very great and a very old one; and when it came first into the Room where the *Prince* was, with a great many

who developed deep research into tropical diseases and Brazilian medicinal herbs. Both, in 1648, published the greatest scientific and artistic work of colonial Brazil: *Historia Naturalis Brasiliae*. This work remains to this day an important source for South American ethnology and ethnolinguistic scholarship, in addition to its evident contributions in the field of natural sciences. For more details, see Willem Piso: *História Natural do Brasil* [1648]. São Paulo 1948.

15 Cf. Evaldo Cabral de Mello: *O Brasil holandês*. São Paulo 2010, p. 241ff.

16 William Temple: Memoirs of what pass'd in Christendom from the war begun in 1672, to the Peace concluded 1679. London 1709.

> *Dutch-men* about him, it said presently, *What a company of White Men are here*? They asked it what he thought that Man was, pointing at the Prince? It answered, *Some General or other*; when they brought it close to him, he asked it, *D'où venez-vous*? it answered, de *Marinnan*. The prince, *A qui êtes-vous?* The Parrot, *A un Portugais*. Prince, *Que fais-tu là?* Parrot, *je garde les poules*. [...] I asked him in what language the Parrot spoke, and he said, in *Brasilian*; I asked whether he understood Brasilian; he said No, but he had taken care to have two Interpreters by him, the one a *Dutchman*, that spoke *Brasilian*, and the other a *Brasilian*, that spoke *Dutch*; that he asked them separately and privately, and both of them agreed in telling him just the same thing that the *Parrot* said. I could not but tell this odd Story, because it is so much out of the way, and from the first hand, and what may pass for a good one; for I dare say this Prince, at least, believed himself in all he told me, having ever passed for a very honest and pious Man; I leave it to Naturalists to reason, and to other Men to believe as they please upon it; however, it is not, perhaps, amiss to relieve or enliven a busy Scene sometimes with such digressions whether to the purpose or no.[17]

From this account, two points are worthy of note: the first, was the fact that the dialogue had been witnessed and confirmed by those who understand both Portuguese and Dutch, in addition to the local governor; and the second was the fact that the tale was narrated as an anecdote to provoke laughter and to divert its readers. Although Locke was not primarily seeking to entertain in *An Essay Concerning Human Understanding*, the episode was recounted as a matter of amusement: it was just a joke in which a parrot reasons, thinks and speaks as if it were a human being. On the strength of such misguided arguments, a parrot could be considered as a human being, just by having the gift of speech. This narrative was reproduced to exhaustion from the end of the seventeenth century onwards;[18] and it became an international joke.[19]

---

17 John Locke: An Essay concerning Human Understanding. Ed. by John Yolton. London 1965, II, ch. 27, § 8.

18 Francis Channing Woodworth: Stories about birds, with pictures to match. Boston 1851. – See also Paul Christopher Johnson: An Atlantic genealogy of "spirit possession". In: Comparative Studies in Society and History 53 (2011), no. 2, p. 393–425; and: Peter Walmsley: Prince Maurice's Rational Parrot: Civil Discourse in Locke's *Essay*. In: Eighteenth-Century Studies 28 (1995), no. 4, p. 413–425.

19 As reported by the diplomat Afonso Franco (cf. Afonso Arinos de Melo Franco: O Índio Brasileiro e a Revolução Francesa. Rio de Janeiro 2000, p. 318).

But was that Locke's real purpose? What did this image represent in the context of the book? If it was assumed that the 'land of parrots', as Brazil was known historically, was linked to the exuberance and beauty of nature, what did it mean that the colourful parrots became a symbol of the country but had an ignoble reputation? At what point in understanding the parrot was there an 'epistemological switch' from a positive interpretation to another, much more negative one? These are themes for examination in the following section.

3 Philosophical Meditations on the Talking Parrot

John Locke used the episode with the parrot as a mere empirical example to highlight one of the greatest philosophical problems of the seventeenth century – and one which remains debated to this day – namely, the definition of humanity or simply the question of personal identity.[20]

At this point, however, it is important to understand the place of this debate within the work itself, which is the heart of *An Essay Concerning Human Understanding*. In book II, Locke challenged Descartes and his famous formulation of *cogito ergo sum* ('I think, therefore I am'), which at that time enjoyed much prestige in English intellectual circles.

What did the French philosopher defend? The existence of innate speculative and practical principles. What did Locke propose instead, after denying there are such innate principles? Locke's great idea was that the origin of all human knowledge lies in experience. In this sense, his central thesis was that people would fill the blank sheet of paper (which is the human mind) with ideas derived from sensation and reflection. According to Locke, human understanding would be completely passive in receiving simple ideas, as basic subjects of all knowledge. Therefore, what humans did not experience – that is, anything which did not arrive mentally through sensation or reflection – would be foreign to human understanding, and thus remain unknown.

---

20 With reference to the currency o Locke's thought, Michael Ayers states: "Perhaps no modern philosopher, except for Kant, has had as broad an influence as Locke, and Kant himself has had a deep debt to his celebrated predecessor. However, Locke has not had a good reception in the 20th century [...]. The *Essay* may be a difficult book to read today, but its reputation is being revived. Philosophers are beginning to relearn how he was able to change the course of European thought, and to find in this book, however dated and problematic as it may be, lessons that are still pertinent to us". (Michael Ayers: Locke. Ideas and Things. São Paulo 2000, p. 10).

The starting point, therefore, of book II, chapter XXVII, in *An Essay Concerning Human Understanding* was precisely the question of identity and diversity. What makes an individual into himself or herself – not someone else? How can individuals retain the same individuality, despite temporal change? Basically, Locke wondered about the criteria by which a person, a human being, or any other animal, is defined. But it must not be forgotten that, for Locke, every word means an idea. Hence to discover what the identity of a certain thing refers to, it is necessary to know what each term means. With the help of this perspective, Locke conceptualized humans within the category of biological animals. The human being was part of the species of *homo sapiens*, whilst simultaneously a person, a legal term, with certain properties of rationality and self-awareness through time. A person, in this context, is a moral, practical term, which denotes the ability to act consciously. Therefore, Locke finishes:

> [...] we must consider what person stands for; which, I think, is a thinking intelligent being, that has reason and reflection, and can consider it self as it self, the same thinking thing in different times and places; which it does only that consciousness, which is inseparable from thinking, and as it seems to me essential to it: it being impossible for any one to perceive, without perceiving, that he does perceive. When we see, hear, smell, taste, feel, meditate, or will any thing, we know that we do so. Thus it is always as to our present sensations and perceptions: and by this every one is to himself, that which he calls self: it not being considered in this case, whether the same self be continued in the same, or divers substances. [...] it is the same self now as it was then; and it is by the same self with this present one that now reflects on it, that that action was done.[21]

Two points are noteworthy in this text by the English philosopher: on the one hand, the person is able to be aware of himself or herself but also, on the other hand, denotes his or her responsibility to others, from the moral point of view. In both cases, humans need to have the thought, the reflection, a sense of what is being thought. That is why there can be no consciousness without the exercise of the mind. Moreover, in defending this argument, the English philosopher established the epistemological basis of the modern conception of personhood, since it was no longer understood as an essence, as Aristotle

21 Locke, Essay, II, ch. 27, § 9.

expounded, or as a substance, as Descartes argued, but as the exercise of mental faculties, purely observable in the empirical sphere.[22]

Thus, when Locke related the human to the parrot, he did not compare them with a prior assumption that the first is rational and that the second is devoid of rational thought. Instead, he approached them equally as biologically organized living beings. For Locke, therefore, the concept of the human as a rational animal is insufficient to explain the difference between the human and the parrot, which was also supposedly intelligent in speech and reasoning. By introducing the subtle difference between a human being and a person, Locke presented the problem of personal identity: what makes each individual into himself or herself.

> For it being the same consciousness that makes a man be himself to himself, personal identity depends on that only, whether it be annexed only too an individual substance, or can be continued in a succession of several substances. For as far as any intelligent being can repeat the idea of any past action with the same consciousness it has of its present thoughts and actions, that it is self to itself now, and so will be the same self as far as the same consciousness can extend to actions past to come; and would be by distance of time, or change of substance, no more two persons than a man be two men, by wearing other cloths today than he did yesterday, with a long or short sleep between: the same consciousness uniting those distant actions into the same person, whatever substances contributed to their production.[23]

Locke thus argued that each person's individual awareness, whether in the past or in the present, according to the actions and thoughts throughout these different epochs, guarantees his or her existence and identity despite changes in time. This approach implied that the self of each one depends upon his or her identity of consciousness. Bernard N. Schumacher summarizes this proposition as follows:

> Locke submits that the criterion for identifying the same person consists on that same person in one interior, and without which one can be founded by an instance outside him. Oneself identity is only accessible

22 Bernard N. Schumacher: La personne comme conscience de soi performante au cœur du débat boéthique. Analyse critique de la position de John Locke. In: Laval théologique et philosophique 64 (2008), no. 3, p. 32.

23 Locke, Essay, II, ch. 27, § 10.

> to oneself, as self-awareness is characterized by a certain authenticity toward oneself.[24]

According to this perspective, all should be aware of their actions and their responsibilities, both present and past, and should also think of possible future actions. Now, none of these undertakings would be possible without the power of memory as a mental process. For Locke, memory[25] constitutes the storage of past ideas. As he argued, it is "the power, of the mind, to often revive additional perceptions of which he has already had them".[26] But, Locke also noted that, memory can fail and, over time, can become slower, hindering the activities of the mind.[27] In sum, Locke wanted to show his readers the beauty, but also the limits of human understanding itself.[28]

Therefore, when evoking the talking parrot, Locke did not need to reinforce his argument that a human, from a biological point of view, is an organic and living being, like a parrot. Instead, he raised the parallel issue of consciousness. Today, 300 years later, it is broadly accepted that animals can be considered as sentient beings, as they feel pain, anguish, loneliness, among other feelings, which is why they may have protected legal status.[29] However, it must be stressed that Locke did not specifically advance this discussion and he left such questions to the naturalists, as already noted.[30]

But if the talking parrot has supposedly rational characteristics, how does it gain its evil characteristics? Why would the Brazilian parrot have the trait of malevolence that would mark its cultural image so negatively? What was the relationship between the representation of the parrot and its wickedness?

In the episode summarised by Locke in *An Essay Concerning Human Understanding*, as recorded earlier, the reader's attention is drawn to a detail which has been little analysed by commentators on Locke: the cleverness of the parrot is so great that it seems to be possessed. Here it is helpful to examine

---

24 Schumacher, La personne, p. 32.

25 For the concept of memory in Locke see Philippe Hamou: Mémoire et 'Conscience continuée': Une lecture de Locke sur l'identité personnelle. In: Philosophical Enquiries: Revue des philosophies anglophones 3 (2014), p. 1–33. On the same topic see also Philippe Hamou: Dans la chambre obscure de l'esprit: John Locke et l'invention du mind. Paris 2018.

26 Locke, Essay, II, 27, § 10.

27 Ibid., II, ch. X, § 8.

28 For this issue see Daniel Silveira: A teoria das ideias e o conceito de identidade pessoal em Locke. Diss. do Programa de Pós-Graduação em Filosofia (PPGF) da Universidade Federal de Sergipe 2020, 100p.

29 Cf. Peter Singer: Questions d'éthique pratique. Paris 1997, 370 p.

30 Locke, Essay, II, ch. 10, § 9.

a further statement within the narrative, when the author wrote: "[…] So that those of his [sc. Nassau's] Train there, generally concluded it [sc. the parrot's skills] to be Witchery or Possession; and one of his Chaplains who lived long afterwards in Holland, would never from that time pardon a Parrot, but said, they all had a Devil in them".[31]

If the parrot's rational trait may seem positive, it does not guarantee how it will be used. Apparently, it was this characteristic of the parrot – its exaggerated expressions, to the point of appearing to be possessed – that historically denoted in Brazilian culture its malevolence. The overall image was therefore one of cleverness, along with a malevolent, biting tongue that defames and mistreats anyone, without a proper ethical control. As a result, those around the parrot responded with a certain fear. That is, the talking parrot is required to provide what it cannot provide: an eventual ethical standpoint, from a cultural point of view. The result is the parrot's systemic lack of virtue.

This relationship between the parrot and its malignancy was not newly determined on the threshold of the eighteenth century; and it did not begin with the story involving Maurice of Nassau. It had already been registered in the previous century and left an inheritance for later times. But where did all this start? Where would the central problem arise?

According to Dante Martins Teixeira, Professor of the Department of Vertebrates, of University Federal do Rio de Janeiro, this problem began when the Christian tradition understood that there were only three classes of beings endowed with words: angels, men and demons. The arrival of the parrot in Europe raised immediate questions. It was true that before the sixteenth century there circulated – still on the European continent – several legends involving demonic animals. Two were particularly remarkable: the first of them dealt with the "Navigations of St. Brandão", which recorded, in a chapter on the "paradise of birds",[32] his arrival on an island of talking birds. By analogy, it might be taken to imply that parrots were nothing but fallen angels who had received soft penalties for not having devoted themselves completely to evil. The second story, also from before the sixteenth century, was narrated by a monk. He observed a parrot, which. when seized by a hawk and realising its imminent danger, expressed itself – still in mid-air – in Latin: "*Sancte Thoma, adjuva me*!" ('Help me, Saint Thomas!').[33] Miraculously, the parrot was released from the hawk's clutches, thanks to the protection of the saint.

---

31 Ibid.

32 Teixeira, p. 102.

33 Ibid.

Yet in the sixteenth century, according to Teixeira, it was Martin Luther[34] who discovered and disclosed the ability of speech and cleverness of the parrot:[35] for him, it was the presence of the devil in the parrot's body which made it imitate people.[36] This assumption encouraged a strong belief that humans should not play with parrots, and that they should be doubly wary of the talking bird. Therefore, says the expert on this subject: "As the most superstitious would still make the sign of the cross in the presence of very wild birds in the second half of the eighteenth century, it does not come as a surprise that, in the year 1792, a parrot and a monkey were processed by heresy and killed at the stake in Miranda del Ebro, Spain".[37]

As such attitudes indicate, Nassau, albeit accompanied by scientists, preferred, in the encounter discussed earlier, to interpret the parrot's behaviour as demonic possession. In a fascinating painting, which is currently located in Krakow, viewers can see the supposed 'true' parrot of the narrative, with a phrase written in Nassau's own hand: "This is the parrot that so skilfully answers and formulates questions that many hundreds of people who heard him could conclude nothing but that it was the demon who spoke through it [...]".[38] Thus, the Dutch governor in Brazilian lands came to the conclusion that, in fact, the parrot was of the devil's party. This story gained ground in such a way that it became popular in Brazil in a more congenial version: it was said that, when confronted with such a smart parrot, someone had the happy idea of sending it to the Dutch king. "His Majesty, however, had never seen such a creature and cut off his head at the beginning of the song, for a talking animal could only be something demonic".[39] In Brazilian popular culture, there continue to be several anecdotes like this one involving the parrot and some pranks committed by it, either with its owners or friends.

The problem of the parrot was not in speaking, much less in imitating the human voice, but that, according to eighteenth-century views. its natural capacity to imitate what was taught to it or simply to express what the parrot

34 Cf. Martin Luther: Werke in Auswahl. Ed. by Otto Clemen. Vol. 8: Tischreden. 3rd ed., Berlin 1962 [1st ed. Bonn 1930], p. 269, no. 5158: "Ego credo Diabolum esse in psittacis vel papagulis, simiis et cercophitecis, quia sic imitari possunt homines".

35 Teixeira, p. 102.

36 Dante Martins Teixeira: Com o diabo no corpo, os terríveis papagaios do Brasil colônia. In: Anais do Museu Paulista 25 (2017), no. 1, p. 87–126; online: https://doi.org/10.1590/1982-02672017v25n0104 [07.12.2020].

37 Ibid., p. 103.

38 Ibid., p. 104. – Cf. also Dante Martins Teixeira (ed.): Brasil Holandês: Miscellanea Cleyeri, Libri Principis & Theatrum rerum naturalium Brasiliae. Rio de Janeiro 1995.

39 Teixeira, Com o diabo no corpo, p. 104.

saw and heard, was a skill without ethical or moral filters. Teixeira provides another example, this time from revolutionary Paris in the 1790s, during the period of The Terror. He records that a parrot shouted non-stop royalist slogans: 'Long live the King!', 'Long live our priests!', 'Long live the nobles!'. The parrot and its owners were arrested and taken before the court. Once there, however, the bird refused to speak in court, and "the defendants would end up being freed for lack of evidence. Nonetheless, the reactionary parrot was forced to be politically indoctrinated by a '*tricoteuse*' – the citizen [Gustave] Le Bon – learning to repeat 'Long live the Nation!' ".[40] It is a story that puts the parrot into the context of laughter, fun, debauchery, irreverence. Hence it seems that, no matter what the culture, the parrot can become inconvenient and therefore maleficent, if it is not given limits: that is, if it is not taught.

4 Conclusion

Finally, Locke's parrot, in addition to the specific issue of *An Essay concerning Human Understanding,* also demonstrates that the relationship between humans and the natural world is difficult and multifaceted, especially with those animals that mimic human traits. It is true that the physiology of the parrot, with its complex syrinx and large brain, provides the animal with an enormous capacity for concentration and learning, quite different from others. This factor causes parrots to imitate the sounds they often hear, either spontaneously or when taught to them. In the case of the Dutchman's story, as related by Locke, he attributed human characteristics to the parrot, on the strength of *its* exceptional powers of speech. Now, for Locke, as has been shown, such an assumption is mistaken. By that argument, anyone hearing a human sound coming out of a computer might immediately conclude that the machine was a human, since it made the same sound. To avoid this misunderstanding, Locke introduced the concept of a person, linked to its legal aspect, and stressed that the definition of a human is accompanied by other attributes, such as consciousness, language, thought.

In Locke, therefore, a human is a living body, but also a person, who is capable of reflection and self-awareness. The parrot, as much as it seems to be the avatar of a human being, because of its cleverness and speech, is still a parrot. In this sense, what distinguishes humans from this animal, as well as from all other non-human animals, is the fact that humans are continuously aware of

40 Ibid., p. 111.

FIGURE 8.1 The devilish parrot (*amazona aestiva*), by Johan Maurits van Nassau-Siegen from *Libri Principis* (1633–1644). The original German handwriting translates thus: "This is the parrot that so expertly answers and asks questions that many hundreds of people who heard it could conclude only that it was the devil that spoke through it. It did not live with me for more than 14 days. When he died, he was as hard as a stick".

participating in life, with affiliations to temporality and memory. It is in this capacity for self-consciousness, finally, that human identity consists: the ability of an individual to recognize himself or herself as being the same person in the present, in the past, and projecting into the future.

Meanwhile, in Brazil, the parrot has the colours of its flag; has a certain exuberance when expressing itself; is traditionally cheerful, like carnival; and yet has abilities that suggest a potential for demonic possession. All these qualities, in one way or another, contributed to the Brazilian people identifying with the parrot, and popularizing jokes involving this talking bird. One of the most famous incarnations, it is said, comes from the end of the story of *Little Red Riding Hood*. Many grandparents found the end of this story sad and, when they told their grandchildren, they provided another version. According to this interpretation, when the wolf knocked on the door, and before Grandma opened it, she heard the voice of the family parrot saying in anguished tones: "Don't open the door! It's the Big Bad Wolf out there".[41] Said and done: Granny was saved, thanks to the ability of the parrot to pay attention to everything that was going on around it. Not fully human, then. But a truly remarkable bird in its own right.

41 This story is narrated by Câmara Cascudo to Audálio Dantas in a meeting between the two. – Cf. Audálio Dantas: Câmara Cascudo e aquela do papagaio. In: Folha de São Paulo; online: https://m.folha.uol.com.br/ilustrissima/2014/03/1425372-camara-cascudo-e-aquela-do-papagaio.shtml [03.06.2021].

CHAPTER 9

Troglodytes, the Monkey Diana and the Aping Swede: Carl Linnaeus on Apes

Annika Windahl Pontén

Résumé

Dans ses travaux de classification zoologique, Carl Linnaeus s'est intéressé à la relation entre les singes et les humains et a retravaillé à plusieurs reprises les définitions et la taxonomie. Son intérêt pour ce sujet s'explique en partie par le fait qu'il s'agissait d'un domaine problématique. Il a également conservé des singes vivants dans le jardin botanique. L'étude de ces animaux a contribué à sa compréhension, à l'évolution des définitions et à la lutte contre la frontière entre l'homme et les autres animaux.

Abstract

In his works on zoological classification, Carl Linnaeus was keenly interested in the relationship between apes and humans and he repeatedly reworked his definitions and taxonomy. His interest in this subject was particularly stimulated by the fact that it was a problematic area. To aid his research, Linneaus also kept live monkeys in his botanical garden. Studying these animals close at hand was very productive. It helped him to refine his definitions and it clarified his efforts to understand the boundary between humans and other animals.

1 Introduction

The Swedish botanist Carl Linnaeus (1707–1778), also known as Carl von Linné, is famous for having classified the three realms of nature (plants, animals and minerals) and as the author of the binary nomenclature for species of plants and animals. It is estimated that he is responsible for the coining of about 5,000 names for species of animals, and about 7,000 species of plants. The realm of plants was in a way his main interest but, since he worked on the whole span of nature and natural history, animals were also a key concern. Thus animals

 | DOI:10.1163/9789004495395_010

were included in his major systematic work entitled *Systema Naturae* (1735),[1] and Linnaeus also produced several dissertations on specific animals.

The 10th edition of the *Systema Naturae,* which appeared in 1758,[2] is of particular interest when it comes to animals, since it is considered to have originated the scientific taxonomy for animals. In fact, animals had been referenced earlier, but there were some important and interesting changes in the 10th edition. Linnaeus included humans among the animals; and it could be said that humans constituted the animal species which Linnaeus studied and analysed the most intensively. He was a professor of medicine, so most of his lectures dealt with various aspects of the human body, disease and dietetics.

In general, he was very interested in the boundaries between different parts of nature, such as the borders between species or the borders between the three realms of nature. This interest was by no means unique to Linnaeus. On the contrary, it was a common topic in natural history at the time.[3] As Robert Wokler has pointed out in his study of Rousseau's views and theories on apes, a number of scientists and naturalists during the eighteenth century "challenged the traditional concept of a decisive intellectual gulf between man and beast".[4] The general idea in Linnaeus's texts was that nature had three realms: animals, plants and minerals. And he examined specific boundaries: between minerals and plants, between plants and animals, but also between humans and other animals.

2 Linnaeus and the Apes

Linnaeus was particularly interested in the similarities between humans and apes. Gunnar Broberg, professor of history of ideas at Lund University, has dedicated much of his research to how Linnaeus studied and defined humans

1 Carolus Linnaeus: Systema Naturæ, sive Regna Tria Naturæ systematice proposita per Classes, Ordines, Genera, & Species. Leiden 1735. – *Systema Naturae* was first published in Holland in 1735 and was consequently published in 13 editions until 1770.

2 Carolus Linnaeus: Systema Naturæ per Regna Tria Naturæ, secundum Classes, Ordines, Genera, Species, Cum Characteribus, Differentiis, Synonymis, Locis. 10th ed., Stockholm 1758.

3 See Silvia Sebastiani: A 'monster with human visage': The Orangutan, Savagery, and the Borders of Humanity in the Global Enlightenment. In: History of the Human Sciences 32 (2019), no. 4, p. 80–99, and Stephen Gaukroger: The Natural and the Human: Science and the Shaping of Modernity, 1739–1841. Oxford 2016, p. 217–264.

4 Robert Wokler: Perfectible Apes in Decadent Cultures: Rousseau's Anthropology Revisited [1978]. In: idem: Rousseau, the Age of Enlightenment, and their Legacies. Ed. by Bryan Garsten. Princeton and Woodstock 2012, p. 1–28, here p. 4.

and, through this research, Broberg has also examined how Linnaeus classified primates and monkeys. Indeed, in his recent biography of Linnaeus, Broberg writes that Linnaeus's interest in primates went much further than did his interest in any other single species, apart from humans.[5]

The dissertation *Antropomorpha* was published in 1760; its thesis was defended by Christian Emanuel Hoppius (b. 1736), but was actually written by his professor Carl Linnaeus.[6] In this work, Linnaeus described three categories of monkeys and apes: those with long tails (monkeys); those with short tails (baboons); and those without tails which he called 'true apes' or 'simia'. In the *Systema Naturae*, Linnaeus had initially named the class of mammals 'Quadrupedia' but that was changed in the 10th edition to 'Mammalia'; and the first order of 'Mammalia' (Quadrupedia), initially called 'Antropomorpha', was renamed as 'Primates'. Linnaeus stressed the similarities between the apes and humans. For instance, he described how they walked in a similar manner and had faces without hair.[7] In the dissertation, he also wrote that: the "genus of apes is very pleasant and much alike us" and he called the apes "the cousins of humans".[8]

There were also creatures about which Linnaeus was uncertain: what were they really? Primates, humans, or something else? Many of them did not exist but he had heard and read stories about fabled creatures, such as those he called 'homo troglodytus' (cave-man) and 'homo caudatus' (tailed man). They were placed at one point in his *Systema Naturae* into a category called 'Monstrosi'; and some of them were also described in the *Antropomorpha* dissertation. Linnaeus further discussed four 'nations' of monkeys and, for all four groups, there were references to other authors, for example to Pliny, Georg Edward and Hans Sloane. The nations were Pygmaeus, Satyrus, Homo caudatus and Troglodyt. The dissertation also included an illustration showing these four groups.[9]

---

5 See Gunnar Broberg: Mannen som ordnade naturen. En biografi över Carl von Linné. Stockholm 2019, p. 329. – Broberg's dissertation discusses Linnaeus's studies and defintions of humans; see Gunnar Broberg: Homo sapiens L.: studier i Carl von Linnés naturuppfattning och människolära. Stockholm 1975 (Lychnosbibliotek Vol. 28).

6 It was the common system that the professor wrote the dissertation and the student defended it at the public defence act in Sweden called *disputation*. There are exceptions where the student wrote the dissertation.

7 See Carl von Linné: Menniskans cousiner (Antropomorpha). Efterskrift och noter av Telemak Fredbärji. Uppsala (Ekenäs) 1955 (Valda avhandlingar Vol. 21).

8 Linné, Menniskans cousiner (Antropomorpha), p. 3–4 (my translation).

9 See Linné, Menniskans cousiner (Antropomorpha), p. 16.

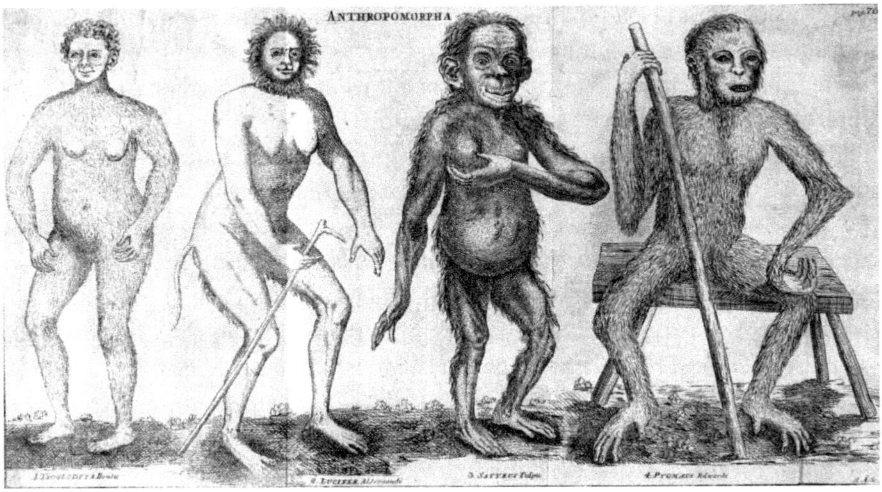

FIGURE 9.1 Christian Emmanuel Hoppius: Homo Troglodytus, Lucifer Aldrovandi, Satyrus Tulpii and Pygmaeus Edwardi (1760)

Among the examples in the *Antropomorpha* dissertation, there were also stories about children who had been brought up among animals and then returned to the human world. Several stories about such children circulated in Linnaeus's time.[10] These children were not necessarily defined as anything else but humans, but they were described as examples moving around or along the boundary between humans and animals. When they first returned to human society, they were more like animals, but the transfer 'back' led to them becoming more and more like humans again. In some respects, they had manifest similarities to apes; and these cases of animal-children also showed the importance of education, climate, and habit (aspects of what scholars today would call 'socialization').

Collectively, the 'wild children', the 'monstrosi' and the apes demonstrated something about the fundamental 'nature' of animals and humans, and pointed to their similarities. Gunnar Broberg has pointed out that the term 'monstrosi' is derived from the Latin word for 'to demonstrate', the idea being that the 'monstrosi' indicated something odd, curious, a hybrid or deformity. The term was not necessarily negative (even if it has often been used in a negative way).[11] The 'monstrosi' were examples of how the borders between categories

10 See Gaukroger, The Natural and the Human, p. 231–232.

11 See Broberg, Mannen som ordnade naturen, p. 333–335.

in Linnaeus's system became problematic and uncertain. That indeterminacy was also one reason why both the terminology and the actual classification changed between the different editions of the *Systema Naturae*. These questions constituted one of the main problems that kept Linnaeus busy through his entire working life.

One of the methods he applied was the study of living monkeys. The first time Linnaeus saw living monkeys was in Holland, in the garden of George Clifford, for whom Linnaeus worked between 1735 and 1738. He then returned to Sweden, where he was appointed professor at Uppsala University in 1741. Not long after, he moved in 1743 into a house in the botanical garden, which is today named after him as the Linnaeus Garden. In the garden, he kept a small menagerie or vivarium, with a variety of birds and animals. There were, for example, exotic birds such as parrots and peacocks, as well as guinea pigs, an agouti, a raccoon and a number of monkeys.[12]

In fact, Linnaeus was not the first professor to keep live exotic animals in the garden. After the garden was founded in 1655, it was developed into one of the largest botanical gardens in Europe. One list of its animals survives from 1685. In 1710 it housed a dromedary, as is indicated by an entry in the garden's account book: "for mending the garden key that the dromedary had broken".[13] The trade in exotic animals was expanding during the eighteenth century, and it became fashionable among wealthy families to keep such animals in private gardens and manorial parks. In Sweden, this trend was apparent across the country; and menageries were kept by people in Carl Linnaeus's network of friends, and at country manors close to Uppsala.

One of the first exotic animals that reached Linnaeus was the North American racoon that he received in the 1740s. This handsome gift came from the Swedish crown prince; and the animal was then named as Sjupp. It is not always possible to establish how the exotic animals were acquired. Most of them were probably gifts, like the raccoon; but some were also purchased. The costs of keeping these live animals became rather large; and in 1760 Linnaeus wrote to the board of the university asking it for more money for this purpose. Linnaeus argued that the exotic animals were beneficial for the whole university; and noted that he had not hitherto asked for any extra money but had paid everything himself. In this letter Linnaeus also noted that most of the animals had been gifts from friends. The board agreed to pay more, especially

12 For a detailed description of the animals kept in the garden see Ingvar Svanberg: "Deras mistande rör mig så hiertligen". Linné och hans sällskapsdjur. In: Svenska Linnésällskapets årsskrift (2007), p. 11–108.

13 Svanberg, "Deras mistande rör mig så hiertligen", p. 37.

FIGURE 9.2 One of the houses for the monkeys which Linnaeus kept in the botanical garden

for the food, granting both additional funds and increased amounts of grain from the university's mill.[14]

Throughout his printed works of synthesis, plus the dissertation and letters, Linnaeus recorded a large number of observations about these animals. When the agouti died, Linnaeus expressed grief and wrote that, after his family, it was the dearest living thing to him. The loss of it really touched his heart.

Linnaeus's observations and descriptions of monkeys were based on the behaviour of the animals to which he had immediate access in the garden. He kept them in wooden houses on poles or pillars (see figure 9.2).

The earliest and most detailed description of a monkey by Linnaeus was published in 1754 in the transactions of the Royal Swedish Academy of Sciences.[15] He wrote that the female monkey, called Diana, ate vegetables but not so much

14 See Svanberg, "Deras mistande rör mig så hiertligen", p. 44–45.

15 See Carl von Linné: Markattan Diana. In: Kungliga Vetenskapsakademiens handlingar. Stockholm 1754, p. 210–217, the illustration p. 239; available online via the Kungliga Vetenskaps Academiens Handlingar database: https://hosting.devo.se/kvah/search.html [01.05.2021].

FIGURE 9.3 Cotton-top tamarin monkey Grinn. Oil painting by Gustav Hesselius [undated]

meat. She loved heat (even if she disliked sitting in sunshine) and had therefore been placed in the heated Orangery. He also described how Diana played with people from the household, particularly the little boys. Yet she could also bite, even if usually not so hard. Diana did not live long. In 1756 Linnaeus wrote to his friend Abraham Bäck that "his beautiful monkey died; where shall I now get a new one?".[16] Apparently, however, a replacement was found. In a list dating from 1769, a monkey from the same species was included, along with six others: *Simia sylvantus, Simia augula, Simia salatrix, Simia sabea, Simia capucina, Simia Oedipus* and *Simia Diana*.[17]

There were also pictures of some of the monkeys, a small number even being painted in oil. A cotton-top tamarin (*Saguinus Oedipus*) called Grinn was painted by Gustav Hesselius. In addition there are two other oil-paintings of other monkeys, and some prints. Other animals were also depicted, for example the racoon and a guinea pig. Even if historians do not know exactly where these paintings were placed in the house, it is clear that the motifs were considered as valuable additions to the decorations. It is possible that they had their place, and/or filled a function, in the parts of the house that were used for lectures.

16 Carl von Linné: Letter to Abraham Bäck. 12th November 1756, Stockholm; online: http://urn.kb.se/resolve?urn=urn:nbn:se:alvin:portal:record-226430 [13.06.2021].

17 See Svanberg, "Deras mistande rör mig så hiertligen", p. 61.

3 Social and Ethical Reflections upon Simian Behaviour

When it came to close observations of the monkeys' behaviour, Linnaeus particularly noted that monkeys easily copy each other: they 'ape'. These observations led to another context where Linnaeus also discussed apes and their behaviour, or rather used them as (anti-)examples. The species was named 'simia' in Latin and the word was also used to denote 'imitation' or 'copying'. Monkeys and their habit of imitating already featured in satires at this time. In particular, they were ridiculed as part of the debates on luxury and conspicuous consumption. There were satirical images showing monkeys as humans full of vanity: for example, dressing up fashionably and preening themselves in front of mirrors. People were also satirized as chattering monkeys, gossiping or spending their time on foppery. A monkey-like behaviour showed that a person was not able to control his or her desires. Such behaviour was considered as passive and imitative rather than active and independent.

All these simian traits were negative, when considered ethically. Linnaeus believed that a person's true self should be reflected on the surface or outside, so that the appearance should mirror who a person really was. If someone imitated others without reflection, the basic connection between the true inner self and the outer surface would be lost. Linnaeus wrote sharply that the royal court resembled a group of apes, who vainly imitated each other. Moreover, he added that one who spends time with apes will become one himself or herself: "Inter simias versantem oportet esse simiam".[18] Linnaeus concluded critically that the Swede of his era was like an ape: a 'Svecus simia', explaining that the Swede: "eats like an Englishman, drinks like a German, dresses like a Frenchman, builds like an Italian, smokes like a Dutchman, takes snuff like a Spaniard, and drinks vodka like a Russian".[19]

This imitation of people from other countries was no good thing, but showed a kind of weakness and lack of independence. There was also a moral risk connected to imitation. When Linnaeus included the human as one of the primates in *Systema Naturae,* he used a Latin motto as a kind of headline: *Nosce te ipsum* – 'Know thyself'. One consequence of imitation may be that a person loses the ability to know his- or herself, and that also meant losing a key part of the quality that Linnaeus defined as human.

---

18 Carl von Linné: Lachesis Naturalis. Uppsala universitets årsskrift, Skrifter med anledning af Linnéfesten. Vol. 11. Ed. by Axel Otto Lindfors. Uppsala 1907, p. 61.

19 Carl von Linné: Diaeta Naturalis (1733). Linnés tankar om ett naturenligt levnadssätt. Ed. by Arvid Hj Uggla. Uppsala 1958, p. 60 (my translation).

Even if the imitation was mainly superficial, it is also worth pointing out that Linnaeus' understanding of the interaction between a person's true, inner self and the outer appearance was more complex. For example, he also used theatrical metaphors and talked about masks and the world as a stage. The use of masks was not always a bad thing, but it took knowledge and experience to master it. Thus a smartly-dressed appearance was not necessarily a result of imitation, or something that deserved critique. Some situations required a particular appearance, or mask, and some people were obliged to appear as finer than others. A person who used masks in a 'correct' or wise way did not merely imitate uncritically. Instead, that person had a character that was visible through a variety of outer appearances.

Questions about the boundaries between different species and various parts of nature may appear as philosophical or metaphysical issues. For Linnaeus, that perspective was probably true to some extent. Yet he viewed the question in practical and pragmatic terms, when it came to explaining how the system worked and how it had developed. Linnaeus' intention was to find and/or to establish a 'natural system' for all species in nature. He was aware that, in the end, his system was not absolutely 'natural' (natural here means that it should reflect the true or objective relationships between the species). Instead, it took pragmatic decisions to settle where the boundary lines should be drawn.

Linnaeus did at times make statements that express a philosophical vision. Thus in his *Philosophia Botanica,* he quoted the Latin maxim that *natura non facit saltum* (i.e. 'nature does not make *discrete* jumps'). However, Gunnar Broberg argues in his most recent biography of Linnaeus, that sentences like these should be understood as citing self-evident axioms, rather than as introducing a hypothesis that needed testing or proving. The texts on systematics and classification were not practical handbooks to agriculture. They did not include many examples concerning the economic use of animals, or discussions of the benefit of animals for humans. So it is probably fair to describe Linnaeus's systematising approach as bridging the concerns of both philosophy and pragmatics.[20]

4 Conclusions: Linnaeus and Human Variety

Finally, it's worth making a couple of remarks on the connection between Linnaeus's descriptions of apes and national characteristics, human varieties

20 See Broberg, Mannen som ordnade naturen, p. 315.

and 'race'. During the spring and summer of 2020, there has been a debate on Linnaeus's role in the history of racism, and there is a connection between that discussion and the themes explored here. This contribution is not written from the perspective of an expert upon theories of racism and anti-racism, but it is based upon a close reading of Linnaeus's views and writings.

Hence it can be said that, when Linnaeus was critical about some types of human behaviours, the notion of imitation was for him a key aspect. He considered that different patterns of behaviour (sometimes imitated) were connected to different groups of people such as 'Swedes', 'Germans', 'sámi-people' or 'Africans'. Linnaeus divided humans into varieties depending on which continent they inhabited (and their skin-colour would have been seen as a consequence of that).

Specifically, in the already mentioned 10th edition of *Systema Naturae* Linnaeus divided the specie *Homo sapiens* into the following four varieties: *Homo americanus, Homo europaeus, Homo asiaticus* and *Homo afer*. Each variety was assigned a temperament from the traditional system of humoral pathology, resulting in descriptions of how each group looks and behaves.[21] The varieties were then put in a list, so there appears to be one variety on top and one at the bottom. However, the top changed from edition to edition. It was either the American (which refers to the indigenous peoples in North America) or the European people who were listed first, but always the Africans who came last.

There have been extensive discussions among scholars as to what extent this listing was 'value-based', But the result is far from clear. On the one hand, the list was not fixed. It changed from edition to edition. On the other hand, it may not be a coincidence that the African was always mentioned last. These descriptions of the different varieties of humans (the term 'race' was not used by Linnaeus) are without doubt stereotypical and could be today considered racist.[22]

However, it is worth stressing that Linnaeus did not have an essentialist view of these humans, with an innate biology prescribed for each group. For example, he considered that bodily properties and social characteristics can change if a person moves to a new terrain, particularly one with a new and different climate. Furthermore, Linnaeus was in fact rather negative and critical about many groups of people and their behaviour. Europeans did not by any means escape his strictures. For example, people from different parts of Sweden were

21 See Linnaeus, Systema Naturae (1758), p. 20–22.

22 See Isabelle Charmantier: Linnaeus and Race; online via the Linnean Society of London, https://www.linnean.org/learning/who-was-linnaeus/linnaeus-and-race [06.11.2020].

described as greedy, stupid, or plain. And others, whether from 'exotic' countries, or from Lapland, were also described in terms associated with the contemporary idea of the 'noble savage'. They were clearly demarcated by their cultural differentness, both in positive and negative ways.

Linnaeus's position on these questions thus turns out to be distinctly complex. And, even if one assumes a direct line from Linnaeus towards a 'modern', 'scientific' racism, that line must be viewed as far from a straight one. There are a number of important differences between Linnaeus concept of 'variety' and the later, more biological ideas of 'race' – ideas which are now themselves very much under challenge.[23]

It is true that Linnaeus's writings and his system of classification have been used in a colonial context and in a racist way by others, both in his own time and afterwards. Yet how other people have used Linnaeus's texts does not in itself clarify Linnaeus' own position. There are many layers within his views, which merit further scholarly examination. Some studies are fruitfully under weigh: an example currently in progress is entitled "Collecting Humanity: Prehistory, race and instructions for 'scientific' travel 1750–1850", led by Linda Andersson Burnett and Bruce Buchan.[24]

Carl Linnaeus had a lifelong interest in how apes and humans were related to each other; and he reworked his definitions and taxonomy many times. His interest in this topic was partly triggered by the fact that it was a problematic field. There were no stable and certain boundaries between different parts of nature or between different species. He thus grappled continually with these definitions.

Meanwhile, Linnaeus's critique of aspects of the behaviour of different groups of humans was more constant. He did not approve of empty imitation, vanity, or any other case when the outer appearance did not correspond to a person's true inner self. He conceived of the primates as 'cousins of men' and saw both important similarities and some differences. Linnaeus studied living monkeys closely, since he had a number of animals in the botanical garden. There are many texts containing observations of the monkeys, and also some pictures of them. Linnaeus seemed to have had an emotional approach to both the monkeys and other animals in the vivarium. Several times, when an

23 See Elazar Barkan: The Retreat of Scientific Racism: Changing Concepts of Race in Britain and the United States between the World Wars. Cambridge 1992; Paul Lawrence Farber: Mixing Races: From Scientific Racism to Modern Evolutionary Ideas. Baltimore 2011.

24 The project has just started and will be continued until 2023. – For more information see https://www.idehist.uu.se/research/research-projects/collecting-humanity/ [06.06.2021].

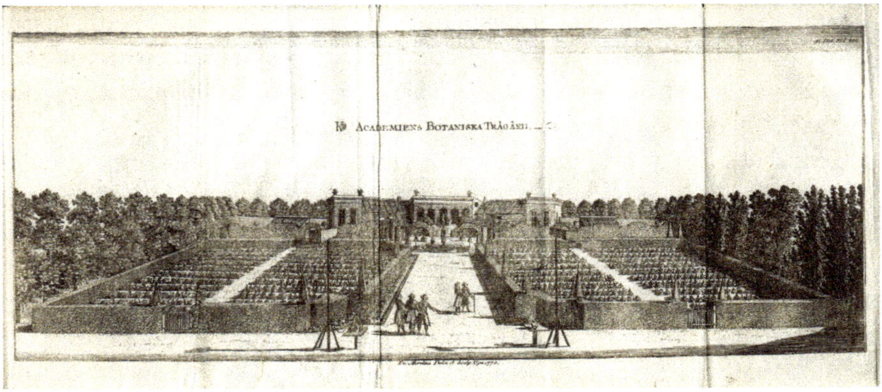

FIGURE 9.4 The botanical garden at Uppsala University by Johann Benedict Busser (1769). The houses for the monkeys are in the front, with 'tables' for the monkeys to sit on.

animal had died, he described them as loved and missed, and almost as family members.

In the 10th edition of his *Systema Naturae* Linnaeus introduced the name 'Mammalia' for the class that earlier had been called 'Quadrupedia'. The first order of that class was changed from 'Antropomorpha' to 'Primates', and the specie *Homo sapiens* (with varieties) and the different apes were included there. In this context, there are a number of problematic issues concerning how humans were defined, described and groups compared to one another. Even if Linnaeus did not use the term 'race', there are aspects of his approach to classification that have been used in racist contexts, both in Linnaeus' time and after.[25] Yet Linnaeus himself was a pragmatist. He followed his practical research, not a pre-set ideology. As already noted, he changed the names and descriptions within his classification system, between different texts and between different editions of *Systema Naturae*. There is thus considerable complexity to be taken into account when analysing Linnaeus's position in this field. Linnaeus' classification has been misused and instrumentalised to justify racist hierarchies and it is therefore even more important to analyse and understand Linnaeus' own work and discuss its legacy.

25 See Linné, Systema Naturae (1758), p. 20–22.

CHAPITRE 10

Les vers à soie et les vers dévoreurs de livres dans une bibliothèque des Lumières. Luxe et morbidité des 'insectes changeants' dans la poésie de Voltaire

Vanessa de Senarclens

Résumé

Cette analyse en deux temps part de l'examen attentif de la matérialité de livres issus d'une bibliothèque démantelée du siècle des Lumières, dont certains ont été transportés après la guerre à Łódź, en Pologne. L'accent est mis d'une part sur les vers à soie – ces 'insectes changeants qui filent la soie' (Voltaire) – et d'autre part sur leurs cousins dévoreurs, les vers de bibliothèque qui creusent des trous dans le papier et laissent ainsi des traces de leur pouvoir destructeur dans les livres. Si les premiers sont des merveilles de la nature, tant louées par les penseurs des Lumières, les seconds visiteurs assidus des ouvrages à l'abandon, sont l'insigne de l'oubli et de la destruction de ce même héritage.

Abstract

This two-pronged analysis starts from a close examination of the materiality of books from a former Enlightenment library (now dismantled) in Łódź in Poland. One focus is upon silkworms – those 'changing insects that spin silk' (Voltaire) – while the other is upon their devouring cousins, the bookworms that leave visible traces of their destructive power in the pages of books. If the former are marvels of nature, much praised by progressive Enlightenment thinkers who advocate their cultivation, the latter are assiduous emblems of oblivion and hence potential destroyers of the same heritage.

1 Introduction

La photo d'un document issu des archives de la Drôme dans la région Rhône-Alpes datant de la fin du dix-septième siècle a récemment fait surface sur les réseaux sociaux[1]. Le manuscrit – sans doute un acte notarié lié à une vente

1 Merci à Dominik Hünniger de m'avoir signalé cette image.

 | DOI:10.1163/9789004495395_011

FIGURE 10.1 Le manuscrit est issu du registre des minutes (années 1673–1676) du notaire Esprit Chabrier de Mirmande des Archives départementales de la Drôme (Région Rhône-Alpes/ France) sous la cote '2 E 14564'

entre particuliers – est illisible: il a perdu sa trame, son papier est perforé de toutes parts, surtout sur les bords qui se sont arrondis jusqu'à en perdre leurs contours. Le manuscrit a circulé sur Twitter avec la légende « Dentelle de vrillette » laissant croire à un ouvrage méticuleux et délicat. Or, la vrillette désigne un insecte xylophage et la dentelle est le produit de ses dévorations du manuscrit. Ce type de document intéresse aujourd'hui moins les historiens des pratiques juridiques de l'Ancien Régime en France que ceux qui réfléchissent aux politiques mémorielles et, plus concrètement, aux stratégies à adopter face aux lentes transformations produites par les vers bibliophages. Les entomologistes recensent de nombreux parasites du papier incluant différentes espèces d'insectes forant les pages des livres et abîmant le cuir des reliures. L'expression du 'ver à livre' s'est imposée notamment en allemand, '*der Bücherwurm*' et, en anglais, '*a bookworm*'[2]. On ne connaît pas d'équivalent exact en français où on se réfère plutôt aux rats pour désigner tant les rongeurs de livres que les grands

2 Alberto Manguel se réfère au ver comme à une personnification du lecteur, cf. Alberto Manguel: The Traveler, the Tower, and the Worm. Philadelphia 2013, p. 89–110 [ch. 3: The Bookworm. The Reader as Inventor of the World].

lecteurs[3]. Le ver forant les livres et les papiers d'archive des étagères à l'abandon est néanmoins aussi en langue français une image parlante de la destruction lente mais inéluctable des documents du passé, un *memento mori* à l'adresse des archivistes, des conservateurs et des historiens.

Cette contribution porte sur les vers dans la poésie du dix-huitième siècle, tant les vers producteurs de soie que ceux perforant le papier et les reliures des livres. Si les vers à soie sont une singularité de la nature que chantent et étudient les poètes comme les savants du siècle des Lumières, les seconds, parasites indésirables sont les emblèmes de la menace sourde qui pèse sur les livres à l'abandon. Partant d'éditions clandestines de poèmes de Voltaire qui se trouvent, depuis l'après-guerre, dans une bibliothèque universitaire polonaise, je m'intéresse tant au ver à soie qu'à son cousin dévoreur de papier. La préposition 'dans' du sous-titre de cet article – Luxe et morbidité des « insectes changeants » dans la poésie de Voltaire – annonce ainsi deux niveaux que je tente d'aborder de front. Cette contribution prend aussi la forme d'un parcours par étapes à la fois dans l'espace et dans le temps, parcours qui suit des volumes clandestins de la poésie de l'auteur des Lumières jusque sur les étagères d'un dépôt de livres déserté.

2 Un combat de vers ?

Commençons par la sommaire présentation des protagonistes de cette lente rixe silencieuse se produisant dans les bibliothèques peu fréquentées: le ver à soie et le ver des livres ont en commun d'être des champions de la mue mais si le premier, associé à la culture du mûrier blanc, a été domestiqué au cours des siècles, le second, sauvage et parasite, prolifère à l'insu des hommes. Les larves des insectes parasites du livre creusent pour se nourrir des petits trous dans le papier qu'elles réduisent à l'état de farine blanche.

A la faveur du perfectionnement des microscopes au cours du dix-huitième siècle, les savants naturalistes découvrirent l'univers « merveilleux » des plus petits animaux de la création auxquels ils vouèrent un intérêt tantôt philosophique, tantôt théologique[4]. La magistrale somme de l'Abbé Noël-Antoine Pluche, *Le Spectacle de la Nature*, parue entre 1732–1742, leur fait une place

3 Cf. Voir sur le sujet Jacques Berchtold: Des Rats et des Ratières. Anamorphoses d'un champ métaphorique de Saint Augustin à Jean Racine. Genève 1992.

4 Cf. les articles réunis dans Nicole Jacques-Chaquin et Sophie Houdard (dir.): Curiosité et libido sciendi. De la Renaissance aux Lumières. Fontenay-aux Roses 2000.

importante[5], mais c'est René-Antoine Ferchault de Réaumur, qui, le premier, leur consacre sept volumes en exclusivité – des chenilles aux scarabées, des vers aux fourmis – dans ses *Mémoires pour servir à l'histoire des insectes* parus entre 1734–1742[6]. Dans son *Dictionnaire raisonné universel d'histoire naturelle* de 1762–1764, Jacques-Christophe Valmont de Bomare précise encore en définissant les vers comme des insectes doubles: tantôt comme des 'merveilles' de la nature lorsqu'ils produisent des fils de soie[7], tantôt comme des « malfaiteurs » lorsqu'ils abiment le bois des grand bateaux transporteurs de marchandises venues de loin: il énumère les « vers rongeurs de vaisseaux » à côté des teignes, des poux et des sauterelles[8].

Du côté des 'merveilles', le ver à soie est déjà bien connu au dix-huitième siècle[9]. Sa chenille secrète un liquide, qui en se durcissant produit un cocon tressé de fils. Il fut découvert et cultivé d'abord en Chine et en Inde pendant l'Antiquité avant d'être importé en Europe, en Italie à la fin du Moyen-âge, puis en France. Au dix-huitième siècle, la sériculture connaît un grand essor en France, surtout dans les milieux protestants qui contribuent à diffuser le savoir de cette industrie dans les pays qui après la révocation de l'édit de Nantes les accueillent, notamment en Prusse[10]. Le ver à soie est amplement

5 Cf. Noël-Antoine Pluche: Le Spectacle de la nature, ou Entretiens sur les particularités de l'histoire naturelle qui ont paru les plus propres à rendre les jeunes gens curieux et à leur former l'esprit. 9 Vols. Paris 1732–1742.

6 Cf. René-Antoine Ferchault de Réaumur: Mémoires pour servir à l'histoire des insectes. 7 Vols. Paris 1734–1742. Sur sa démarche scientifique et son esthétique, cf. Jean-Christophe Abramovici: Du « Merveilleux vrai » des « petits animaux ». Réaumur, entre rococo et Lumières. Dans: Dix-huitième siècle 42 (2010), no. 1, p. 305–320.

7 Cf. Jacques-Christophe Valmont de Bomare: Dictionnaire raisonné universel d'histoire naturelle: contenant l'histoire des animaux, des végétaux et des minéraux [...]. Paris 1765. – Ici selon: idem: [Art.] 'ARA'. Nouvelle édition, revue & considérablement augmentée par l'auteur. Paris 1775, t. 1, p. 280, il précise: « On aura sans doute regret, dit M. de Réaumur, de ce qu'il nous reste si peu d'espérance de profiter d'une découverte si ingénieuse. Mais, ajoute-t-il, il reste encore quelque espèce de ressources: peut-être trouvera-t-on des araignées qui donnent plus de soie que celle que nous voyons communément dans le Royaume. Il est certain, par le rapport de tous les voyageurs, comme nous le verrons plus bas, que celles de l'Amérique sont beaucoup plus grosses que les nôtres, d'où il semble qu'elles doivent faire de plus grosses coques. Les vers à soie, quoiqu'originaires des pays éloignés, nous aideroient même à espérer que les araignées de l'Amérique pourroient vivre dans ceux-ci. Quoi qu'il en soit, il faut expérimenter: c'est la seule voie de découvrir des choses utiles & curieuses ».

8 Bomare, Dictionnaire raisonné universel d'histoire naturelle, p. XIII–XIV.

9 Cf. l'ouvrage recent de Ben Marsh: Unravelled Dreams. Silk and the Atlantic World 1500–1840. Cambridge 2020.

10 Le roi Frédéric II a largement contribué au développement de la culture du mûrier blanc et des manufactures de soie en Prusse, cf. à ce sujet le catalogue de l'exposition, Susanne Evers (dir.): Seiden in den preußischen Schlössern: Ausstattungstextilien und Posamente unter Friedrich II. (1740–1786). Berlin 2014.

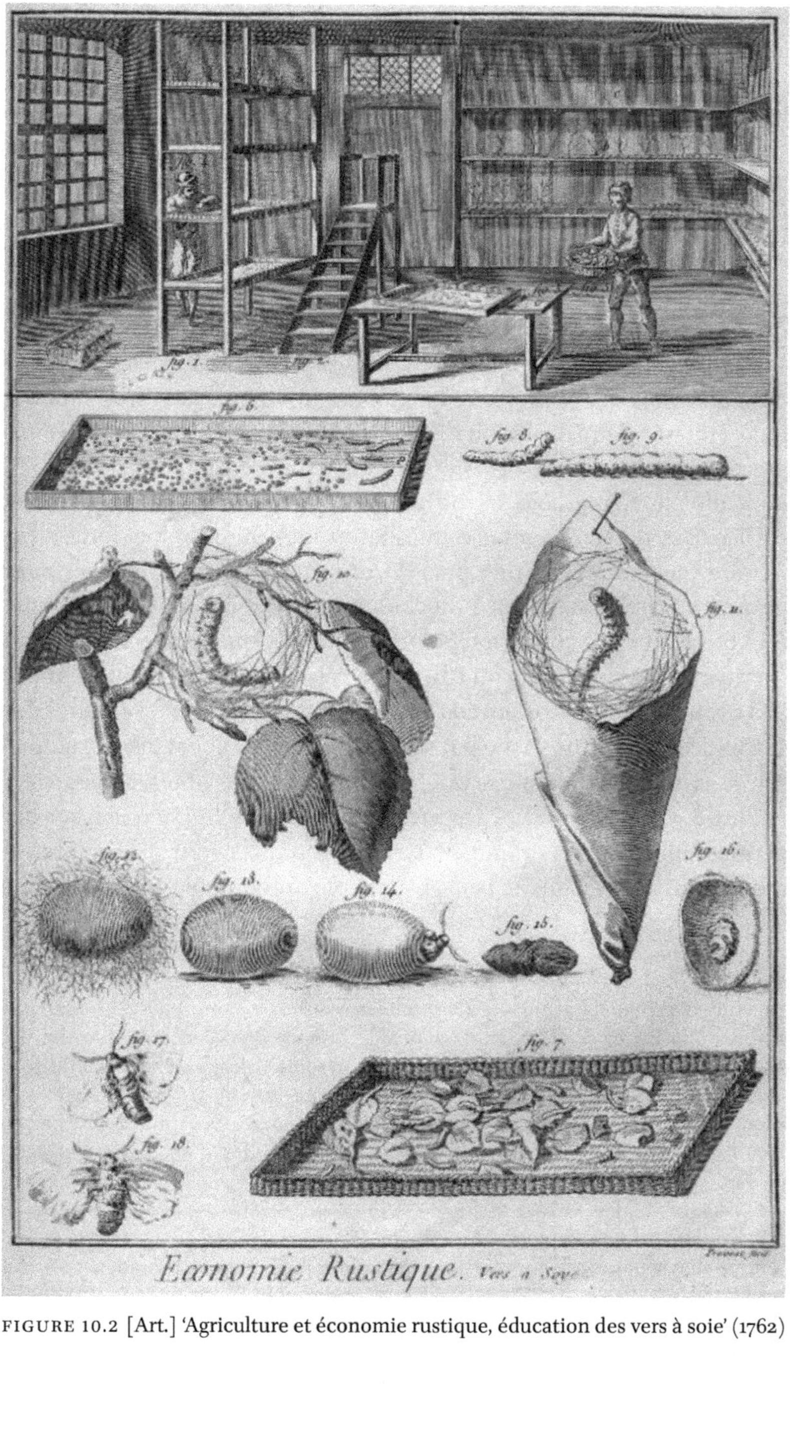

FIGURE 10.2 [Art.] 'Agriculture et économie rustique, éducation des vers à soie' (1762)

décrit dans l'*Encyclopédie ou Dictionnaire raisonné des sciences, des arts et des métiers* comme dans ses Planches qui se réfèrent aux études microscopiques de Marcello Malpighi (1628–1694) et d'Antoni van Leeuwenhoek (1632–1723) un siècle plus tôt: « Ver-à-soie, (*Science microscop.*) le *ver-à-soie* est un animal dont chaque partie, soit dans l'état de *ver,* soit dans celui de mouche, mérite quelques regards [...] ». Et comme la légende de la planche (*Ver à Soye: Oeconomie rustisque*) l'indique, cet insecte intéresse tant les naturalistes que les économistes soucieux du développement agricole français. Il est même question dans la légende de l'image de « l'éducation des vers à soie »[11].

Du côté du ver 'malfaiteur', il faut attendre le dix-huitième siècle pour qu'il soit clairement identifié et répertorié. L'*Histoire abrégée des insectes qui se trouvent aux environs de Paris* (Paris, 1762) d'Etienne-Louis Geoffroy propose le terme de vrillette pour ce ver qu'il examine tout à son affaire. Il l'observe perçant le bois et creusant « des trous ronds, comme ferait une vrille »[12]. Poursuivant plus loin l'enquête au moyen d'un microscope plus performant, le suédois Charles de Geer propose une description morphologique de ces insectes guère plus grands qu'une puce[13]. Au tome 4 de ses *Mémoires pour servir l'histoire des insectes*, parus à Stockholm en 1774, il décrit la « vrillette pointillée » ou anobie ponctuée (*anobium punctatum*) capable de passer « par les plus petites fentes des tiroirs où l'on garde ces choses, & perçant même le bois de ces tiroirs de petits trous ronds qui donnent passage ou les reliures des livres »[14]. Charles de Geer ne cache pas son admiration pour cet insecte auquel il prête un « naturel surprenant », une « opiniâtreté » et une « constance » remarquables. Le « sang-froid » des vrillettes lui rappelle l'héroïsme « des Sauvages prisonniers de l'Amérique »[15]. Comme eux, elles font surtout preuve de discipline dans l'adversité. Et leur arme la plus redoutable est la capacité

11 Agriculture et économie rustique, éducation des vers à soie. Dans: Robert Morrissey et Glenn Roe (dir.): Encyclopédie, ou dictionnaire raisonné des sciences, des arts et des métiers [...]. Denis Diderot et Jean le Rond d'Alembert (dir.). Chicago 2017; en ligne: https://artflsrv03.uchicago.edu/philologic4/encyclopedie1117/navigate/18/31/ [08.01.2021].

12 Etienne-Louis Geoffroy: Histoire abrégée des insectes qui se trouvent aux environs de Paris. Dans laquelle ces Animaux sont rangés suivant un ordre méthodique. Paris 1762, t. 1, p. 109.

13 Cf. M. le Baron Charles de Geer: Mémoires pour servir l'histoire des insectes. Stockholm 1774. T. 4, p. 227.

14 « Comme elles sont petites, elles parviennent à s'introduire partout, passant par les plus petites fentes des tiroirs où on garde ces choses, &perçant même le bois de ces tiroirs de petits trous ronds qui leur donnent passage, de sorte qu'il est très difficile de s'en garantir », Geer, Mémoires pour servir l'histoire des insectes, t. 4, p. 234.

15 Geer, Mémoires pour servir l'histoire des insectes, t. 4, p. 228–229.

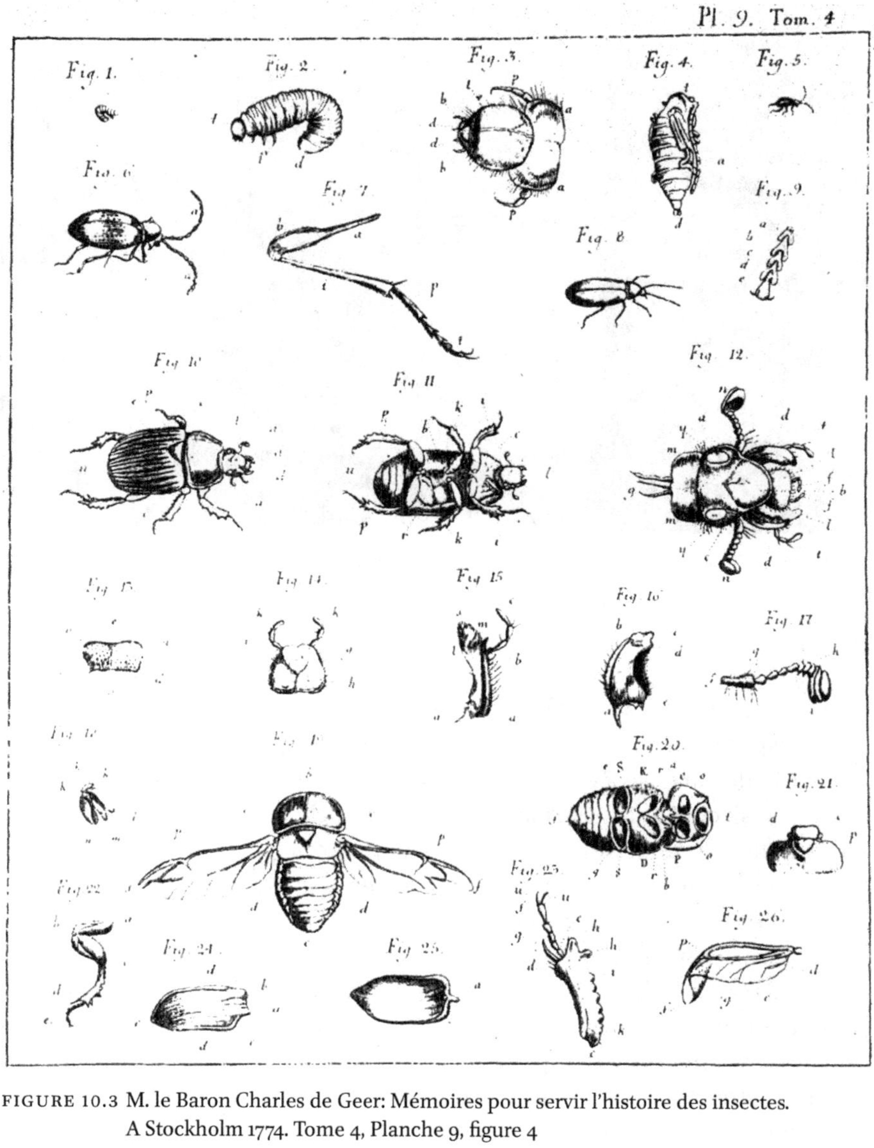

FIGURE 10.3 M. le Baron Charles de Geer: Mémoires pour servir l'histoire des insectes. A Stockholm 1774. Tome 4, Planche 9, figure 4

à passer inaperçues. Cette qualité de discrétion n'est guère sollicitée dans le biotope où elles sévissent, dont il est question maintenant.

Il est un lieu où le travail opiniâtre des vers rongeurs de livres s'opère à l'abri des regards, et auquel même les habitués des bibliothèques n'ont pas accès. Il s'agit de la salle de dépôt des anciens imprimés ('*Stare druki*') de la Bibliothèque

FIGURE 10.4 Photo du dépôt des anciens imprimés de la Bibliothèque Universitaire de Łódź comme champ de bataille méconnu et ignoré entre les vers 'merveilles' et 'les vers malfaiteurs'

Universitaire de Łódź en Pologne ('*Biblioteka uniwersytecka Łodzi*') où ils ont été transportés après la Deuxième Guerre mondiale. Cette grande ville au sud-ouest de Varsovie a été un haut lieu de l'industrie textile au cours du dix-neuvième siècle. On la désignait comme la 'Manchester de l'Est' car cette ville industrielle exportait ses vastes productions principalement dans l'Empire russe. Son université fut inaugurée en janvier 1945 alors que Varsovie, entièrement démolie par les années de terreur nazie, était à terre. Des stocks de livres, issus des bibliothèques d'anciennes provinces allemandes ont constitué les premiers fonds de cette bibliothèque. Ils sont arrivés à Łódź dès 1946–1947, d'abord entreposés dans une usine désaffectée où l'humidité et les parasites ont laissé des traces de leur passage[16]. À partir des années 1950, les livres ont été enregistrés et ont reçu une cote d'entrée. En grande majorité, ils sont issus d'anciennes bibliothèques allemandes privées, ecclésiastiques et publiques, en l'occurrence de la Bibliothèque d'État prussienne et de la Bibliothèque

16 Cf. Dorota Bartnik et Tomasz Piestrzyński: Zwischen Tradition und Modernität. Zur Geschichte der Universitätsbibliothek in Łódź. Dans: Cora Dietl et Malgorzata Kubisiak (dir.): Unbekannte Schätze. Germanica des 16. Jahrhunderts in der Universitätsbibliothek Łódź. Łódź 2018, p. 29–46, ici p. 30.

municipale de Berlin[17]. Ces deux dernières institutions avaient commencés, dès 1941 à évacuer leurs fonds de livres et de manuscrits en dehors de la ville de Berlin, choissisant des régions rurales peu peuplées en Poméranie orientale et en Silésie afin de les abriter des bombardements[18]. Le 19 janvier 1960 – date anniversaire de la libération de la ville de Łódź de l'occupation allemande – après quatre ans de construction, une nouvelle bibliothèque fut inauguré à Łódź avec une grande salle de lecture et une plus petite pour les anciennes collections. Non loin de celle-ci, au bout d'un corridor se trouve le dépôt des livres anciens. C'est une salle rectangulaire avec des livres empilés jusqu'au plafond, organisés en étagères et en galeries reliées par des échelles. Les livres ne sont pas en accès libre mais doivent être commandés pour être consultés.

L'un des ouvrages se trouvant dans ce dépôt illustre parfaitement le combat silencieux et presque ironique des vers dans le temps. C'est un livre de jardinage du dix-huitième siècle, *Neue Garten-Lust* réimprimé à Leipzig en 1742, avec des annotations et un exlibris renvoyant à une bibliothèque de Poméranie[19] qui conseille le lecteur comment se mettre à l'abri des plantes et des animaux nuisibles. On y voit d'un côté un homme chapeauté occupé à redresser les branches d'un arbre fruitier et à organiser son jardin, et de l'autre, les tunnels forés par les dévorations des vrillettes. Comme si les formations organiques et tortueuses du vers dans le papier venait contrarier l'élégante planification du jardinier.

3 'Pièces fugitives' et celles disparues

On trouve dans cette même salle des éditions rares du dix-huitième siècle, des ouvrages qui portent les sigles d'anciens propriétaires, pour la plupart d'origine allemande. Appréhendés comme objet dans leur dimension matérielle, ces livres estampilés et souvent annotés font resurgir les divers contextes

17 Pour les collections privées, voir Tomazs Ososiński: Peter Kameke und seine Bibliothek. Dans: Dietl et Kubisiak, Unbekannte Schätze, p. 147–163; Vanessa de Senarclens: Teile einer verstreuten Büchersammlung aus dem 18. Jahrhundert. Die Bibliothek Schloss Plathe und ihre Benutzer. Dans: Dietl et Kubisiak, Unbekannte Schätze, p. 117–136.

18 Cf. Jakub Gortat: Drucke des 16. Jahrhundert aus der Preußischen Staatsbibliothek in der Universitätsbibliothek Łódź. Dans: Dietl et Kubisiak, Unbekannte Schätze, p. 97–108.

19 L'ouvrage contient la signature d'un lecteur ainsi qu'une date (1749) mais aussi un *ex libris* de la Bibliothek Schloss Plathe dont il sera question plus tard: Heinrich Hessens et Theodorus Phytologus: Neue Garten-Lust: das ist, Gründliche Vorstellung, Wie ein Lust-Küchen- und Baum-Garten unter unserem Teutschen Climate füglich anzurichten: allerhand sowohl fremde als einheimische Blumen, [...]. Leipzig 1742 [1690].

FIGURES 10.5–10.6 Page de garde du volume Heinrich Hessens, Theodorus Phytologus: Neue Garten-Lust: das ist, Gründliche Vorstellung, Wie ein Lust- Küchen- und Baum-Garten unter unserem Teutschen Climate füglich anzurichten: allerhand sowohl fremde als einheimische Blumen, [...]. Leipzig 1742.

historiques qu'ils ont traversés et esquissent, au hasard d'une annotation des marges, le profil d'un lecteur ou d'une lectrice[20]. Dans le lot de ces livres à Łódź, il existe plusieurs éditions de l'œuvre poétique de Voltaire, certaines éditions pirates de poèmes jugés scandaleux pour leur libertinage ou leur impiété, mais aussi d'autres mélanges reliés par des libraires avisés souvent à l'insu de l'auteur sous le titre de *Pièces fugitives*. Ce sont des ouvrages interdits que les libraires font circuler sous le manteau par l'intermédiaire de réseaux bien organisés dans toute l'Europe[21].

Les volumes en français de Voltaire dans le dépôt de la bibliothèque universitaire de Łódź sont issus dans leur grande majorité d'une bibliothèque fondée au milieu du dix-huitième siècle, à la lisière du royaume de Prusse dans le château d'une petite ville de Poméranie orientale nommée Plathe. Elle fut fondée par un savant et sa femme qui collectionnaient les ouvrages des philosophes

20 Cf. Vanessa de Senarclens: Verlegt, verwahrt und vergessen. Die Bücher aus den ehemaligen deutschen Bibliotheken in Polen. Dans: Merkur. Deutsche Zeitschrift für europäisches Denken 74 (2020), no. 857, p. 77–84.

21 Cf. à ce sujet les différents ouvrages de Martin Mulsow, en particulier Martin Mulsow: Prekäres Wissen. Eine andere Ideengeschichte der Frühen Neuzeit. Berlin 2012.

des Lumières et possédaient de nombreux livres en français[22]. Après la prise de Plathe par l'Armée rouge le 4 mars 1945, la bibliothèque a été placée sous tutelle soviétique jusqu'en 1947 et certains de ses livres ont été transportés en Union Soviétique. Mais, suite aux accords de Potsdam du 2 août 1945, la plupart des livres de cette bibliothèque de la ville devenue polonaise sous le nom de Płoty – le nombre est estimé à environ 13 000[23] – a été attribuée par le gouvernement polonais à la Bibliothèque Universitaire de Łódź (BUŁ)[24].

Le syntagme 'poésie fugitive' désigne au dix-huitième siècle des pièces de circonstances, souvent brèves, d'une inspiration philosophique sensualiste. Dans l'ouvrage qu'elle consacre à ce genre, Nicole Masson rappelle la définition donnée dans l'*Encyclopédie*: « On appelle pièces fugitives, tous ces petits ouvrages sérieux ou légers qui s'échappent de la plume et du porte-feuille d'un auteur, en différentes circonstance de sa vie, dont le public jouit d'abord en manuscrit, qui se perdent quelquefois … »[25]. Ces vers de circonstance, sérieux ou badins, sont marqués par le modèle de la conversation philosophique et imprégnés d'une esthétique rococo célébrant le moment de plaisir et la fugacité du temps[26].

On trouve dans le département des livres anciens de la bibliothèque universitaire de Łódź plusieurs volumes de 'pièces fugitives' de Voltaire, parus de son vivant. Ce sont des specimens intéressants de l'édition clandestine ou semi-clandestine des Lumières. Par exemple un *Recueil de Nouvelles pièces fugitives de Mr. de Voltaire* paru à Genève en 1759, mais aussi des mélanges de différents textes à l'initiative d'un libraire qui se garde bien de donner son nom. Une brochure cartonnée gris-bleu est classée sous le nom de '*Crébillon fils*' avec la cote '*1006412*' dans le catalogue de Łódź. Le recueil s'ouvre sur un conte oriental codé consacré aux amours du roi Louis XV et porte le titre *Les amours de Zeokinizul, roi des Korirans, ouvrage traduit de l'Arabe du voyageur Krinelbol* (À Amsterdam 1756).

22 Cf. Vanessa de Senarclens: Spinoza in Hinterpommern: Ein wiederaufgetauchter Bibliothekskatalog von 1756 in kulturwissenschaftlicher Perspektive. Dans: Das achtzehnte Jahrhundert 4444 (2020), no. 1., p. 51–65.

23 Bartnik et Piestrzyński, Zwischen Tradition und Modernität, p. 29.

24 Ibid., p. 32.

25 Nicole Masson: La Poésie fugitive au XVIIIe siècle. Paris 2002, p. 21.

26 Kim Gladu: La grandeur des petits genres. L'esthétique rococo à l'âge de la galanterie, Paris 2019, p. 10; mais aussi Vanessa de Senarclens: Über ein erotisches Gedicht Friedrichs des Großen, seine kürzliche Wiederentdeckung und die Querelle der Gelehrten. Versuch einer Einordung der flüchtigen und sinnlichen Verse des Preußenmonarchen in ihren literaturgeschichtlichen Kontext. Dans: Germanisch-Romanische Monatsschrift 63 (2013), no. 2, p. 261–268.

En sus de ce pamphlet satirique et donc risqué pour les libraires, il contient aussi plusieurs poèmes de Voltaire, notamment son *Poème sur le désastre de Lisbonne*, écrit suite au tremblement de terre de Lisbonne, en 1756 et un *Poème sur la loi naturelle*, antérieur, dont la dernière strophe est une 'Prière' en huit vers adressée à Dieu. Cette déclaration de foi a été ajoutée dans l'édition en question, une démarche qui implique une connaissance intime de l'œuvre de Voltaire [voir Illustration 8, 9 et]. On trouve aussi à Łódź des *Mélanges de poésie, de littérature, d'histoire et de philosophie*, parus au tome 2 de la *Collection complette des œuvres de M. de Voltaire.* [Genève, Cramer], 1756. 17 tomes. 8°. Il s'agit cette fois des premières éditions autorisées par l'auteur parues de son vivant.

Certains titres de l'œuvre poétique de Voltaire qui se trouvaient jusqu'en 1945 dans la bibliothèque de Poméranie ne sont pas arrivés jusqu'à Łódź: au cours des grands chambardements de la fin de la Deuxième guerre mondiale, ils ont sans doute été égarés, volés, pilés ou détruits. Des témoins des événements rapportent, qu'à la fin de l'hiver 1945, les livres servaient de matière isolante dans la recontruction des murs ou de buches pour se chauffer. Ces œuvres disparues seraient aujourd'hui des sources intéressantes tant sur le contexte clandestin de leurs impressions que sur celui de leur diffusion jusqu'aux confins de la Poméranie orientale au dix-huitième siècle. Dans le lot des livres disparus de Voltaire, on peut citer deux titres en particulier: *Le Porte-Feuille trouvé, ou Tablettes d'un curieux*, paru à Genève (Chez les Libraires associés) 1759 et un *Recueil de pièces fugitives en vers, contenant: l'épitre à Uranie. L'épitre à Athénais. Question de théologie, avec la réponce* (sic) *et la réplique. L'Anti-theologien et la Bathsébath*, paru à Londres (Chez Jean-Pierre Schmidt) en 1744. Le nom de l'éditeur comme le lieu d'impression faussent les pistes. On connaît le *Recueil de pièces fugitives en prose et en vers, par M. de V****, édité en 2012 par Olivier Ferret et Myrtille Méricam-Bourdet, mais il a un autre contenu[27]. Les catalogues de la bibliothèque de Plathe, tant le catalogue manuscrit de 1756[28] que celui de 1900[29] contenaient une attribution intéressante deux épîtres voisines

27 Voltaire: Recueil de pièces fugitives en prose et en vers, par M. de V***. Olivier Ferret et Myrtille Méricam-Bourdet (dir.). Saint-Étienne 2012. Sur le poème principal du recueil, voir Nicholas Cronk: L'Épître à Uranie de Voltaire: stratégies de publication d'une œuvre clandestine. Dans: Béatrice Guion: Poétique de la pensée. Études sur l'Âge classique et le siècle philosophique en hommage à J. Dagen. Paris 2006. p. 275–284.

28 Catalogus Bibliothecae secundum materias in sectiones descriptae, Platea ad Regam, 1756: Manuscrit se trouvant dans les archives Johann Carl Conrad Oelrichs de la Staatsbibliothek zu Berlin sous la cote SBB-PK, Abt. IIIa acc. Mas 1978.

29 Les scans ont été mis en ligne, voir http://www.bismarck-osten.com/zettelkatalog [14.12.2020], III P Belletristik: Voltaire.

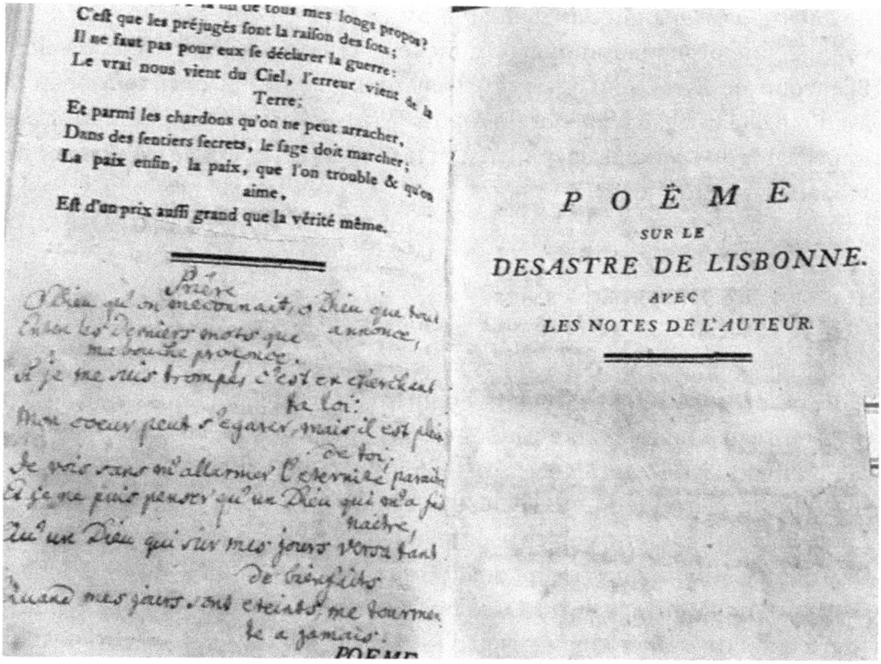
C'est que les préjugés sont la raison des sots;
Il ne faut pas pour eux se déclarer la guerre:
Le vrai nous vient du Ciel, l'erreur vient de la Terre;
Et parmi les chardons qu'on ne peut arracher,
Dans des sentiers secrets, le sage doit marcher;
La paix enfin, la paix, que l'on trouble & qu'on aime,
Est d'un prix aussi grand que la vérité même.

POËME
SUR LE
DESASTRE DE LISBONNE.
AVEC
LES NOTES DE L'AUTEUR.

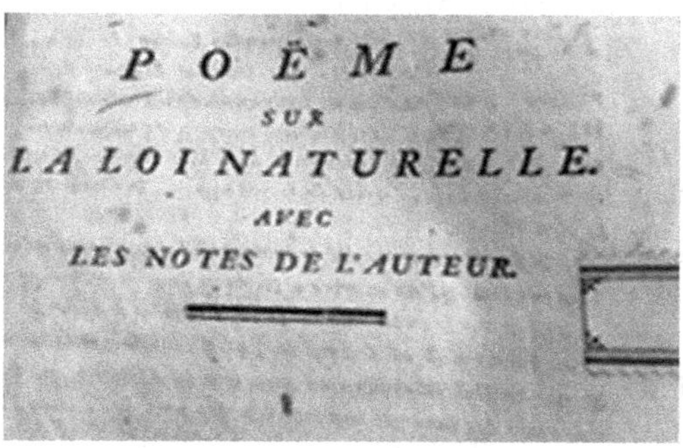
POËME
SUR
LA LOI NATURELLE.
AVEC
LES NOTES DE L'AUTEUR.

FIGURES 10.7–10.8 Recueil anoyme de plusieurs ouvrages réunis sous la cote '1006412', sous le titre de la première œuvre de manière anonyme sous le titre *Les amours de Zeokinizul, roi des Korirans, ouvrage traduit de l'Arabe du voyageur Krinelbol*, à Amsterdam 1746. Le recueil comprend également deux poèmes de Voltaire annotés: le *Poëme sur la loi naturelle* et le *Poëme sur le désastre de Lisbonne, avec les notes de de l'auteur*.

de l'*Epître à Uranie* qui désignaient le roi de Prusse comme co-auteur de ce volume, une piste amusante mais difficile à vérifier sans le livre en question. Beaucoup de livres sont issus de la même bibliothèque dont le fondateur avait été chambellan à la cour du roi de Prusse, entre 1745 et 1749. Il n'est pas impensable que certains documents sur la production poétique clandestine de Frédéric II s'y trouvaient.

4 Faire des vers sur les vers

À l'entrée 'VER. n. masc'. du *Gradus français, ou Dictionnaire de la langue poétique* paru en 1822 à Paris, l'auteur énumère les motifs poétiques associés au « vermisseau » rampant. Il puise ses exemples dans les œuvres des poètes du dix-huitième siècle français et cite pêle-mêle Jacques Delille, Voltaire, François-Joachim de Pierre de Bernis, Louis-Pierre-Marie-François Baour-Lormian et Louis Racine. En tant qu'objet poétique, le ver semble pris dans un faisceau d'attributs antinomiques, tantôt qualifié de « vil », d'« impur », d'« immonde » dans les « replis tortueux » où il « se traîne », tantôt de « précieux », d'« industrieux », ou de « laborieux » lorsqu'il est associé à la culture de la soie « d'Ispahan » et à ses horizons exotiques et luxueux[30]. Ainsi en poésie, le vers évoque ainsi tant les gadoues locales qu'une Perse reluisante qui fait rêver.

Avec ses exemples, le *Gradus français* ne s'y trompe pas: le ver fait une entrée fracassante dans le domaine poétique à une époque qui s'intéresse philosophiquement aux plus petits animaux de la nature. Dans leurs études sur la poésie du dix-huitième siècle, Sylvain Menant, Michel Delon et Catriona Seth ont aussi chacun montré la porosité entre les découvertes scientifiques et la poésie des Lumières[31]. Le microscope et les longues vues des savants ouvrent sur de nouveaux arpents du réel. Les poètes renouvellent leurs sujets en consacrant des odes aux théories nouvelles de Newton, de Kepler, mais aussi de Buffon, de Réaumur et de Linné. Le vaccin contre la petite vérole est l'occasion d'une ode: *L'inoculation* par Claude-Joseph Dorat – « Que la raison de l'homme,

---

30 Louis J. M. Carpentier: Gradus français, ou Dictionnaire de la langue poétique précédé d'un nouveau traité de la versification et suivi d'un nouveau dictionnaire des rimes. Paris 1822, p. 1133.

31 Cf. Sylvain Menant: La Chute d'Icare. La Crise de la poésie française (1700–1750). Genève 1981 (Histoire des idées et critique littéraire Vol. 193). – Cf. aussi les anthologies de Catriona Seth (dir.): Anthologie de la poésie française: XVIIIe siècle. Paris 2000 (Bibliothèque de la Pléiade Vol. 467); Michel Delon (dir.): Anthologie de la poésie française du XVIIIe siècle. Paris 1997 (Collection poésie Vol. 316).

incertaine et captive,/ S'affranchit lentement du joug qui la captive! »[32] On fait aussi des rimes sur la loi de la gravitation et on insère des calculs algébriques dans des épîtres amoureuses – « Mais, hélas! A + D – B/ N'est pas = à je vous aime »[33]. On chante aussi les 'merveilles' d'une nature redécouverte par la science. Pierre Fulcrand de Rosset loue dans son poème didactique intitulé *L'Agriculture*, au chant III, tant l'admirable capacité de travail du ver que celle de l'industrie française qui sait en faire bon usage[34]. Il s'inscrit dans la tradition des *Géorgiques* de Virgile[35] en proposant de longues descriptions instructives sur la culture de vers à soie. « Du fruit de ses travaux voulez-vous profiter? », s'enquiert-t-il, recommandant aux cultivateurs les mesures propres à dompter la molle nature par des activités de production utiles: « Le fil est détaché, roule & vous obéit;/ Votre main le gouverne, en ordre il déploie,/ Se forme en écheveau, & vous donne de la soie »[36]. Comme celle de Rosset, la poésie descriptive de Jean-François de Saint-Lambert loue moins des laborieux insectes que l'homme ingénieux qui, par son savoir et son intelligence, sait mettre à profit les travaux des insectes rampants. Dans le *Discours préliminaire* à son poème *Les Saisons* (Amsterdam, 1769), il théorise une poésie renouvelée par les nouvelles approches scientifiques[37].

Les insectes font également leur entrée dans la poésie de Voltaire. Les sept poèmes qui constituent le *Discours en vers sur l'homme* paraissent entre 1734 et 1737, d'abord de manière anonyme dans divers recueils de pièces fugitives[38]. Abrité par la clandestinité, Voltaire y formule une morale humaine émancipée de la religion chrétienne. Le quatrième discours intitulé *De la modération en tout, dans l'étude, dans l'ambition, dans les plaisirs* interpelle le savant mu par

32 Seth, Anthologie de la poésie française, p. 208.

33 Cf. les vers de Voltaire à Madame du Châtelet « lorsqu'elle apprenait l'algèbre ». Dans: Sylvain Menant (dir.): Œuvres de 1734–1753. Œuvres complètes de Voltaire. Vol. 14. Oxford 1989, p. 517.

34 Pierre Fulcrand de Rosset: L'Agriculture. Poëme. Paris 1774, p. 110.

35 Voir son *Discours sur la poësie géorgique* en Introduction.

36 Rosset, L'Agriculture, p. 111: « Du fruit de ses travaux voulez-vous profiter?/ Avant que son réveil puisse vous arrêter, Dépouillez les rameaux; qu'une chaleur puissante/ Étouffe sous son toit la nymphe languissante./ Alors de ses tissus que l'eau tiède amollit,/ Le fil est détaché, roule & vous obéit;/ Votre main le gouverne, en ordre il déploie,/ Se forme en écheveau, & vous donne de la soie. »

37 « La Philosophie a pour ainsi dire agrandi et embelli l'univers, [...]. Les progrès des sciences comprises sous le nom de Physique, l'Astronomie, la Chimie, l'Histoire naturelle, etc. ont fait connaître le palais du monde, et les hommes qui l'habitent [...]; on a pu entreprendre des poëmes qui demandent une connoissance variée de la Nature. » (Jean-François de Saint-Lambert: Les Saisons: Poëme. Amsterdam 1769, Discours préliminaire, p. XII.

38 Voltaire: Discours en vers sur l'homme. Dans: Haydn T. Mason (dir.): Œuvres complètes de Voltaire. Vol. 17. Oxford 1991, p. 389–531.

« l'amour de la science » qui, foin de systèmes interprétatifs préconçus, sait observer la nature: « La nature est ton livre, et tu prétends y voir/ Moins ce qu'on a pensé que ce qu'il faut savoir »[39]. Son *Discours en vers sur l'homme* salue notamment les travaux de Réaumur sur les insectes qui éclairent un ordre naturel irréductible à tout système interprétatif clos :

> Réaumur dont la main si savante et si sûre
> A percé tant de fois la nuit de la nature,
> M'apprendra-t-il jamais par quels subtils ressorts
> L'éternel Artisan fait végéter les corps ?[40]

En matière de science, Voltaire s'en prend aux bâtisseurs de systèmes, aux dogmatiques et aux adeptes de systèmes interprétatifs préconçus[41]. Dans la nature, il cherche un ordre divin, une loi universelle. Dans ce même *Discours*, il évoque aussi les vers qui partagent avec l'homme la charge d'être des créatures naturelles, à la fois pathétiques et grandioses :

> D'où vient qu'avec cent pieds, qui semblent inutiles,
> Cet insecte tremblant traîne ses pas débiles ?
> Pourquoi ce ver changeant se bâtit un tombeau,
> S'enterre, et ressuscite avec un corps nouveau,
> Et le front couronné, tout brillant d'étincelles,
> S'élance dans les airs en déployant ses ailes ?[42]

L'enchantement de Voltaire pour les métamorphoses du vers devenu papillon apparaît dans cet extrait mais aussi dans de nombreux textes en prose, tels les *Éléments de la Philosophie de Newton* de 1736, l'article 'Métamorphose, métempsychose' du *Dictionnaire Philosophique* de 1764 ou les *Singularités de la Nature* de 1768[43].

---

39 Voltaire, Discours en vers sur l'homme, p. 492.

40 Ibid., p. 493.

41 Voir l'étude importante de Gerhardt Stenger: Voltaire 'naturaliste'. Dans: idem (dir.): Œuvres complètes de Voltaire, Vol. 65B. Oxford 2017, p. 1–100. – Stenger souligne l'immense culture scientifique de Voltaire et rappelle que tant à Cirey dans sa retraite des années 1730 chez Mme du Châtelet que plus tard à Ferney, Voltaire ne s'en tient pas seulement à des connaissances livresques mais expérimente lui-même avec des limaçons ou rassemble des observations géologiques: « Jusqu'à la fin de sa vie, Voltaire a mené le combat contre l'arrogance du dogmatisme et l'esprit de système », p. 14.

42 Voltaire, Discours en vers sur l'homme, p. 493.

43 Voir sur le sujet l'article 'Métamorphose, métempsycose' dans le *Dictionnaire Philosophique*. Dans: Christiane Mervaud (dir.): Œuvres complètes de Voltaire. Vol. 36.

Son *Poème sur la loi naturelle* évoque également les insectes rampants dans le contexte d'une réflexion sur l'"Être suprême', ordonnateur du monde. Publié pour la première fois en 1756, il date en fait des années prussiennes (1750–1752)[44] et manifeste un déisme s'élaborant en réponse au matérialisme de Julien Offray de La Mettrie et Paul Henri Thiry d'Holbach répandu à la cour de Frédéric II. Le poème s'ouvre par un exorde où le poète annonce chercher Dieu, non dans le « labyrinthe obscur de la théologie », ni dans ces « romans qu'on appelle systèmes », mais bien dans une nature, qui « en sait plus qu'ils n'en ont jamais dit »[45]. Le titre de la deuxième partie, *Réponses aux objections contre les principes d'une morale universelle*, formule une exigence morale. Les « insectes changeants qui nous filent la soie »[46] sont partie intégrante d'une démonstration philosophique. Ils ont une place au sein de l'ordre de la nature, au même titre que les abeilles et leur miel.

> Les insectes changeants qui nous filent la soie,
> Les essaims bourdonnants de ces filles du ciel
> Qui pétrissent la cire et composent le miel,
> Sitôt qu'ils sont éclos forment-ils leur ouvrage ?
> Tout mûrit par le temps, et s'accroît par l'usage.
> Chaque être a son objet, et dans l'instant marqué
> Il marche vers le but par le ciel indiqué[47].

Au chapitre 7 de son essai de synthèse sur l'histoire ancienne *La Philosophie de l'Histoire* de 1765, Voltaire revient sur les lois de la nature. Il discerne un « instinct » de développement basé sur des transformations incessantes à l'œuvre dans la nature mais aussi dans l'histoire et cite à l'appui de cette thèse le passage du *Poème sur la loi naturelle* relatif à l'abeille qui fait le miel et au ver qui produit du fil à soie[48].

Oxford 1994: « N'est-il pas bien naturel que toutes les métamorphoses dont la terre est couverte, aient fait imaginer dans l'Orient où on a imaginé tout, que nos âmes passaient d'un corps à un autre; un point presque imperceptible devient un ver, ce ver devient papillon; un gland se transforme en chêne, un œuf en oiseau; l'eau devient nuage et tonnerre; le bois se change en feu et en cendre; tout paraît* enfin métamorphosé dans la nature. » p. 370.

44 Voltaire: Le Poème sur la loi naturelle. Dans: Haydn T. Mason et Thomas Wynn (dir.): Œuvres complètes de Voltaire. Vol. 32B. Oxford 2007, voir introduction : p. i–xv.

45 Voltaire: « Exorde », dans: idem, Le Poème sur la loi naturelle, p. 58.

46 Ibid., p. 64.

47 Ibid.

48 Voltaire: La Philosophie de l'Histoire. Dans: John H. Brumfitt (dir.): Œuvres complètes de Voltaire. Vol. 59. Genève 1969, p. 111.

5 Conclusion: Plantons des mûriers!

Les vers à soie incarnent pour Voltaire, et ce dès le début de sa carrière de poète, le travail de la civilisation basé sur les conquêtes de l'intelligence humaine, de la raison et des capacités des individus et des nations à collaborer plutôt que de se détruire par les guerres ou de s'autodétruire par manque de tolérance. Le poème *Le Mondain* de 1736, au début de sa carrière, associe la soie au « superflu, chose très nécessaire » et rend grâce à « l'âge d'or » du présent: « Regrettera qui veut le bon vieux temps, [...]. Moi, je rends grâce à la nature sage/ Qui, pour mon bien, m'a fait naître en cet âge »[49]. Cet optimisme quant à un projet de paix et de prospérité basé sur des connaissances nouvelles et fondant des liens de commerce entre les nations esquissait les contours d'un avenir prometteur. Quelques années plus tard, les écrits de Voltaire sur l'histoire de France (*Le Siècle de Louis XIV,* 1751 puis *L'Essai sur les Mœurs et l'esprit des nations,* 1756) portent eux un regard plus nuancé sur le passé et plus inquiet quant à l'avenir. Ils soulignent chacun la dimension civilisatrice de la culture des vers à soie au dix-septième siècle et déplorent les dévastations causées à cette industrie par les guerres de religion en France. La révocation de l'Édit de Nantes qui accordait des droits aux réformés sous Louis XIV réduit à néant les efforts concertés de générations; la France se vide des protestants spécialisés dans la soie qui émigrent à l'étranger et emportent avec eux leur savoir-faire.

Un dialogue fictif paru à l'entrée 'Droit de guerre' de l'œuvre alphabétique tardive intitulée *Questions sur l'Encyclopédie pour un amateur*, 1771, rappelle, à la fin de la carrière de Voltaire, le point de vue d'un moraliste imaginaire. On croit entendre la voix de l'auteur. Il dit son accablement au spectacle incessant des violences, des guerres et des barbaries, il rêve aussi tout haut d'un prince « juste et bienfaisant » qui saurait rétablir un contexte de confiance propice au bonheur de la nation :

> Misérables mortels, soyez justes et bienfaisants, cultivez la terre et ne l'ensanglantez pas. Princes, n'allez pas dévaster l'héritage d'autrui, de peur qu'on ne vous tue dans le vôtre ; restez chez vous, pauvres gentillâtres, rétablissez votre masure ; tirez de vos fonds le double de ce que vous en tiriez ; entourez vos champs de haies vives ; plantez des mûriers ; que vos sœurs vous fassent des bas de soie ; améliorez vos vignes[50].

49 Seth, Anthologie de la poésie française, p. 75.

50 Voltaire: [Art.] 'Droit de Guerre'. Questions sur l'Encyclopédie par des amateurs. Dans: Christiane Mervaud et Nicholas Cronk (dir.): Œuvres complètes de Voltaire, Vol. 40. Oxford 2018, p. 581.

La formule « plantez des mûriers » lancée aux misérables mortels ressemble à celle, plus fameuse, engageant à « cultiver notre jardin » qui clôture le fameux conte philosophique *Candide ou de l'optimisme* de 1759[51]. Dans la correspondance de la fin de sa vie, Voltaire renvoie souvent à la culture du vers à soie, à la fois comme une métaphore de son combat d'homme des Lumières, mais également comme investissement concret dont il sait bénéficier et dont il fait bénéficier[52]. Rappelons qu'il vit alors depuis des décennies en exil dans sa résidence de Ferney aux portes de Genève, d'où il intervient régulièrement dans les affaires politiques françaises pour prêter main forte par ses écrits aux victimes de l'intolérance religieuse. Plusieurs affaires judiciaires mobilisent son énergie à réhabiliter les innocents, la première et la plus fameuse concernant le protestant Jean Calas, accusé à tort de la mort de son fils et mis à mort sur la roue sur la base de préjugés religieux à l'origine du *Traité sur la Tolérance* de 1763. Voltaire entretient aussi à Ferney des manufactures d'horlogerie et de soie où il attire de nombreux artisans. Parmi eux, Pierre Calas, le fils de l'innocent que Voltaire avait réussi à faire réhabiliter de manière *post mortem*. Deux lettres datant de septembre 1769 témoignent de ses démarches de manufacturier de la soie et de l'importance économique mais également symbolique de cette activité pour celui qu'on appelait déjà le patriarche de Ferney. La première est adressée à la duchesse de Choiseul à laquelle il vante les prodiges produits par ses cultures de vers à soie et dans laquelle il mentionne incidemment le rôle qu'il y joue le fils Calas :

> [...] mes vers à *soie* qui m'ont donné *de* quoi faire ces *bas*, ce sont mes mains qui ont travaillé à les fabriquer chez moi avec le fils *de* Calas ; ce sont les premiers *bas* qu'on ait faits dans le pays. Daignez les mettre madame une seule fois, montrez ensuite vos jambes à qui vous voudrez, et si on n'avoue pas que ma *soie* est plus forte et plus belle que celle *de* Provence et d'Italie je renonce au métier[53].

51 Voltaire: Candide ou l'optimisme traduit de l'allemand de Mr. le docteur Ralph. Dans: René Pomeau (dir.): Voltaire. Romans et contes. Paris 1966, p. 259.

52 « Les vraies richesses sont chez nous; elles sont dans notre industrie. Je vois cela *de* mes yeux. Mon blé nourrit tous mes domestiques; mon mauvais vin qui n'est point malfaisant les abreuve; mes vers à *soie* me donnent des *bas*; mes abeilles me fournissent d'excellent miel et *de* la cire; mon chanvre et mon lin me fournissent du linge ». Dans: Voltaire: Correspondence and related documents. Dans: Theodore Besterman (dir.): The complete works of Voltaire. Vol. 120. Oxford 1975 [Lettre à Pierre Samuel Du Pont de Nemours, 16 July 1770 (D16525)], p. 337–338.

53 Ibid. [Lettre à Louise Honorine Crozat Du Châtel, duchesse de Choiseul, 4 septembre 1769 (D15869)], p. 208.

Dans une lettre écrite quelques jours plus tard, cette fois à la marquise du Deffand, Voltaire évoque la joie, les distractions et la fierté que lui procurent ses cultures de vers à soie dans son domaine au pied du Jura: « J'ai trouvé, lui écrit-il, le secret d'avoir des vers à soie dans un pays tout couvert de neiges, sept mois de l'année, et ma soie, dans mon climat barbare, est meilleure que celle d'Italie »[54]. Dans un contexte politiquement décourageant pour celui qui, dans cette même lettre, se définit comme « un vieux solitaire » cette formule tient de la bravade et du défi.

54 Ibid. [Lettre à Marie de Vichy de Chamrond, marquise du Deffand, 6 septembre 1769 (D15876)], p. 217.

CHAPTER 11

Electoral Animals in Eighteenth-Century England

Matthew O. Grenby and Kendra Packham

Résumé

La présente contribution examine quelques-unes des façons dont les animaux ont été mis en vedette dans le processus électoral et le discours électoral en Angleterre tout au long du XVIIIe siècle. En s'appuyant sur une série de documents textuels et visuels, la contribution défend la thèse que l'utilisation généralisée de l'imagerie animale pour représenter les participants humains aux élections (notamment les candidats au Parlement et les 'hordes' électorales) montre comment le vote public était interprété et considéré dans une société encore incertaine de la valeur et de la viabilité de la démocratie.

Abstract

This essay examines some of the ways in which animals featured in the electoral process, and electoral discourse, in England in the long eighteenth century. Drawing on a range of textual and visual materials, the essay argues that the widespread use of animal imagery to portray human participants in elections (whether parliamentary candidates or in electoral 'crowds') comments on how public voting was understood and regarded in a society still unsure of the value and viability of democracy.

1 Introduction

Domestic animals have become a noticeable feature of recent elections, at least in the United Kingdom and United States. In the 2019 UK General Election, the press, social media and the public were captivated by photographs of dogs at polling stations, usually seen waiting outside while their owners (among them the Prime Minister Boris Johnson) cast their votes.[1] In the US 2020 presidential

1 See e.g. Dogs at polling stations: Pooches at the polls. BBC News, 12 December 2019; online: https://www.bbc.co.uk/news/uk-50755053 [26.10.2020]. The wider phenomenon has its own Wikipedia article: https://en.wikipedia.org/wiki/Dogs_at_polling_stations [26.10.2020].

 | DOI:10.1163/9789004495395_012

election, attention focused on a handful of incidents of pets, even dogs and cats long dead, being registered as voters – examples of electoral fraud, according to the incumbent President Donald Trump, which could call the election result into question.[2] More metaphorically, the 'dead cat' has become an established political tactic, used to describe the act of deliberately introducing a shocking statement into a political campaign so that reporters and the public immediately overlook every other issue: as if (as Boris Johnson explained while Mayor of London) a dead cat had been thrown on a dining table.[3]

But animals taking a prominent role in elections is far from new. This essay seeks to open up the subject, taking a first look at some of the ways in which animals featured in the electoral process and discourse in England in the long eighteenth century.[4] Whether alive or dead, a range of nonhuman animals were involved, exploited and often maltreated in elections, as in other aspects of eighteenth-century life.[5] Animals were used as modes of transport and conveyance for candidates and voters for example, or to provide the basic materials of electoral practices like quills, parchment and textiles. Or they were on the menu at election entertainments and dinners.

However, this essay focuses instead on how animals were represented in the discourse that grew up around elections, conveyed through pamphlets and prints, ballads and reportage, and many other forms of cultural production. These texts and images, artefacts and practices, were often intended as interventions in individual electoral contests. But the place of animal imagery

2 See Warren joked about her dog voting Democrat. GOP officials insist electoral pet fraud is no laughing matter. In: Washington Post, 17 July 2020; online: https://www.washingtonpost.com/nation/2020/07/17/dogs-voting-biden-trump/ [29.10.2020]. – The 2020 US presidential election also featured the Twitter hashtag #Dogs4Biden.

3 See Boris Johnson: This cap on bankers' bonuses is like a dead cat – pure distraction. In: The Daily Telegraph, 3 March 2013; online: https://www.telegraph.co.uk/news/worldnews/europe/eu/9906445/This-cap-on-bankers-bonuses-is-like-a-dead-cat-pure-distraction.html [29.10.2020]. – See also Sam Delaney: How Lynton Crosby (and a dead cat) won the election. In: The Guardian, 20 January 2016; online: https://www.theguardian.com/politics/2016/jan/20/lynton-crosby-and-dead-cat-won-election-conservatives-labour-intellectually-lazy [29.10.2020].

4 The fullest analysis of animal imagery in British political culture so far is Christopher Machell: The Politics of Bestial Imagery in Satire, 1789–1820. [unpublished PhD thesis, Northumbria University 2011]; online: http://nrl.northumbria.ac.uk/id/eprint/5846/1/machell.christopher_phd.pdf [29.10.2020].

5 On the different experience of animals, animal cruelty, and the development of an animal protection movement and literature of compassion for animals in the eighteenth century, see Kathryn Shevelow: For the Love of Animals. The Rise of the Animal Protection Movement. New York 2008; Jane Spencer: Writing about Animals in the Age of Revolution. Oxford 2020.

within electoral discourse also comments more generally on how public voting was understood and regarded in a society still unsure about the value and viability of democracy.

2 Animal Identities for Candidates in Contested Elections

Parliamentary elections in eighteenth-century Britain determined which candidates were selected to take seats in the House of Commons, one of the two houses forming the British parliament. Only a small proportion of the population were entitled to vote. Voters were, in theory, exclusively men over twenty-one who either resided, held property, or were a member of a trade guild in a particular constituency. Borough constituencies represented towns and cities, including some that had formerly been major centres, but over centuries had declined in importance and population: the so-called 'rotten' boroughs. County constituencies represented rural populations as well as those voters in urban centres which were not counted as boroughs (including many of the new industrial cities). In some boroughs, the franchise was relatively wide, including almost all adult men. In others, the electorate could be limited to a few, usually those owning or renting a certain kind of property. The system allowed for a high degree of control by those with wealth or local power, especially since votes were cast in public. In fact, in most constituencies, actual elections seldom took place, since local magnates could in effect select their chosen candidates, or because, in the more open and populous constituencies, the cost of influencing the electorate was so high that candidates declined to contest an election. This haphazard system remained in place, largely untouched, until a major Reform Act of 1832 which expanded the electorate and set a precedent for further reforms. Even during the long eighteenth century, however, elections were not just for an elite. When electoral contests took place they were often surrounded by a upsurge of public debate and activity lasting weeks or months. This state of affairs affected and engaged a wide section of the population – men and women, adults and children, rich and poor, franchised and unenfranchised – all of whom participated as consumers and often active participants in this periodic political carnival.[6]

---

6 Studies of eighteenth-century electoral culture and political participation include Frank O'Gorman: Voters, Patrons, and Parties. The Unreformed Electoral System of Hanoverian England, 1734–1832. Oxford 1989; Frank O'Gorman: Campaign Rituals and Ceremonies. The Social Meaning of Elections in England, 1780–1860. In: Past & Present 135 (1992), p. 79–115; Mark Knights: Representation and Misrepresentation in Later Stuart Britain. Partisanship and Political Culture. Oxford 2005, ch. 4. On the electoral involvement of women, see Elaine

Animal imagery was a familiar part of electoral discourse. In particular, would-be Members of Parliament were often presented as animals of various kinds in the graphic prints, broadsides and ballads published during election campaigns. Frank O'Gorman notes some spectacular examples related to the Coventry election of 1780. Coventry had one of the largest electorates of any borough and was controlled by the Whig party dominated by Protestant Dissenters ('the Yellows'). When their oligarchy was challenged by Anglican, Tory candidates ('the Blues') a vicious campaign saw the publication of numerous broadsides (large sheets of paper, bearing cheaply printed text and images on one side only, designed to be handed round or pasted on walls). These included *A large and curious Collection of Beasts, Birds, Reptiles, &c.* depicting the leading Blues as animals: an ape, a serpent, and servile dogs. A second broadside introduced a crocodile, jackdaw and a chameleon, personifying one of the leading Blues (the text explained) because it was "an Animal that changes Colour at Pleasure".[7] Satirical political prints (engraved images with captions, sometimes hand-coloured and aimed chiefly but not exclusively at affluent and literate audiences) similarly presented candidates as animals, or sometimes strange hybrid creatures not unlike mythological beasts.[8] Figure 11.1 shows James Gillray's 1807 depiction of the election hustings in Covent Garden, London where the candidates to become MP for Westminster are scrambling up a pole (a visual metaphor for the 'poll', or election). At the top is an enormous goose, identifiable from his human head as Sir Francis Burdett who had won the election. From his prominent beak issues a goose's hiss directed towards the crown and constitution: Gillray's strike against Burdett's radicalism. (Note also the dead cat on the roof of the hustings, an electoral animal to which we will return in due course.)[9]

Chalus: Elite Women in English Political Life c.1754–1790. Oxford 2005. – On the importance of literature to electoral discourse and participation, see Kendra Packham: Literature and the Culture of Elections and Electioneering in Eighteenth-Century England. In: The Review of English Studies 2020; online: https://doi.org/10.1093/res/hgaa049 [29.10.2020]. – The AHRC-funded Eighteenth-Century Political Participation and Electoral Culture project (ECPPEC) (Principal Investigator: Matthew Grenby, Co-Investigators: Elaine Chalus and Tom Schofield, Research Associates: Kendra Packham, James Harris, Dan Foster-Smith and John Schoneboom) will provide fresh insights into participation in pre-Reform elections. The project website will be online at https://ecppec.ncl.ac.uk/.

7 Frank O'Gorman: Coventry Electoral Broadsides, 1780. In: The Yale University Library Gazette 67 (1993), no. 3/4, p. 161–169, here p. 168.

8 See Eirwen E. C. Nicholson: Consumers and Spectators. The Public of the Political Print in Eighteenth-Century England. In: History 81 (1996), no. 261, p. 5–21.

9 James Gillray: *Election – candidates; – or – the republican-goose at the top of the pol(l)e* (1807); online: https://www.britishmuseum.org/collection/object/P_1851-0901-1230 [26.10.2020]. (Alternatively: https://www.npg.org.uk/collections/search/portrait/mw62772/Election-candidates---or---the-Republican-goose-at-the-top-of-the-pole [29.10.2020].) – See also Gillray's

FIGURE 11.1 James Gillray: *Election – candidates; – or – the republican-goose at the top of the pol(l)e*. London 1807.

Triumphal procession of little-Paul-the taylor upon his new-goose (1806); online: https://www.britishmuseum.org/collection/object/P_1851-0901-1218 [26.10.2020]; alternatively: https://www.npg.org.uk/collections/search/portrait/mw62753/Triumphal-procession-of-little-Paul-the-taylor-upon-his-new-goose [26.10.2020].

Dogs, cats, livestock and poultry were of course very familiar to eighteenth-century city dwellers, as well as the rural population. Popular culture was full of animals too, and the zoomorphisation of candidates could draw upon these forms. *Bow Wow Wow*, for example, was a popular ballad extolling the virtues of dogs, frequently reprinted following what may have been its first appearance in the 1786 John O'Keeffe comic opera *Patrick in Prussia, or, Love in a Camp*.[10] By 1790, it had been adapted as 'The New Bow Wow', a satirical commentary on the personalities of the Newcastle-under-Lyme election:

> The Burgess he's an honest dog, he'll vote for old friends, sir,
> The Admiral he's a staunch dog, he'll never run away, sir;
> The candidate he's a slippery dog, he'll change as the wind, sir,
> And if that you wou'd know his name, 'tis buckram and tape, sir;
> *Bow wow wow, fal lalderiddy iddy, bow wow wow*.[11]

The symbolic deployment of animal imagery would also have been familiar from fables and other kinds of allegorical texts and imagery. Associating an individual politician with a particular animal guaranteed instant recognition, even for non-literate audiences, and opened up many possibilities for satire. Gillray, for instance, consistently presented Burdett as a goose in a succession of caricatures, perhaps in reference to the occasion in the 1806 Westminster election when supporters of Richard Brinsley Sheridan had paraded a tableau of a tailor, tape measure, shears, cabbage and a goose to the hustings in mockery of James Paull, a tailor's son, whom Burdett was then supporting.[12] Gillray lampooned Paull's victory in *The Triumphal procession of little-Paul-the Taylor upon his new-goose*, with Paull, sitting on a cabbage and holding the tape measure and shears, riding a large goose with Burdett's face.[13] The next year, as they prepared for a further Westminster election, Paull and Burdett fell out, and Gillray caricatured their duel in *The Patriot's deciding a point of Honor!*, the

10 See the notation of Bow wow wow: https://tunearch.org/wiki/Annotation:Bow_Wow_Wow [26.10.2020].

11 In: Hannah Barker and David Vincent (ed.): Language, Print and Electoral Politics 1790–1832. Newcastle-under-Lyme Broadsides. Woodbridge 2001, p. 18. Buckram is a kind of fine linen or cotton fabric, probably meant to denote the trading (as opposed to land-owning) connections of the candidates being attacked here.

12 See Marc Baer: The Rise and Fall of Radical Westminster, 1780–1890. Basingstoke 2012, p. 109.

13 See Gillray, *Triumphal procession of little-Paul-the Taylor upon his new-goose*.

FIGURE 11.2 [Anon.]: *Westminster Election. 1780. A Devil looking over Lin*[*col*]*n*. London 1780.

goose (now labelled 'Sir Francis Goose') exchanging pistol shots with Paull, still distinguished by his tailor's scissors (the injuries they sustained are still visible in figure 11.1, published three weeks later).[14]

In fact, the politician-as-animal trope had probably been at its apogee slightly earlier, centring on Charles James Fox, who had been a Member of Parliament for Westminster from 1780 until his death in 1806, when Paull and Burdett strove to take his place. There were two reasons for this. First, Westminster, with its large electorate formed of every male homeowner, was a crucial and hotly contested constituency for both the Whig and Tory parties. Second, Fox's name immediately suggested his animal identity. There is evidence that effigies of foxes were paraded by his supporters. In a print satirically depicting the hustings at the 1780 Westminster election for example (figure 11.2), a stuffed fox has been mounted on a long pole, with a label reading "Reynard for Ever". In the foreground a man wearing a hat in the shape of a fox (labelled "Volunteers Fox") leads a procession of supporters who also wear brushes, being distributed by a

14 See James Gillray: *Patriot's decinding a point of Honor! 1807*; online: https://www.britishmuseum.org/collection/object/P_J-3-60 [26.10.2020].

woman at the right of the image – although whether these are all real fox tails (customarily called 'brushes') or merely brooms designed to resemble them is unclear. Political prints certainly cannot be taken as straightforward evidence for electoral behaviour.[15] Yet these depictions have a ring of truth (and other prints corroborate the parading of stuffed foxes mounted on poles), as does the presence of barking dogs adding to the chaos, and another dead dog (or cat) which has apparently been flung through the air (both at the front left of the scene).[16]

Less credible, obviously, is the fox sitting in the lap of Britannia, who stands behind Fox on the dais, with her traditional emblem of a lion at her feet, and the depiction of the election clerks, sitting below the candidates, as monkeys. Fox's opponents cheerfully represented him in vulpine form too, as on a 1796 election token (bronze medals commemorating the result) bearing the image of a rampant fox with the legend "Some of the foxes tricks on a Westminster poll", or a 1790 satirical print in which beer-drinking men dance round a fox hanging from a gibbet that has been set on fire.[17] Often his supporters were represented as geese (perhaps another source for Burdett's representation in that form). The accusation was, of course, that, as in a fable, the cunning fox had duped the foolish fowls. As Harriet Guest has noted, a 1784 election ballad *The Fox and Geese, or, a late House of Commons in Westminster* explains how the fox changes his principles to suit any situation, and several prints show Fox either leading or beguiling a crowd of geese.[18]

15 See Ludmilla Jordanova: Image Matters. In: The Historical Journal 51 (2008), no. 3, p. 784–785.

16 Cf. Robert Dighton's pen and watercolour drawing of the Westminster election of 1788, discussed in Penelope J. Corfield: "For I will Consider my Cat Jeoffry". Cats and Creativity in Eighteenth-Century Britain, in the present volume (see figure 12.4, p. 178); or Thomas Rowlandson's *Procession to the Hustings after a Successful Canvass* (1784); online: https://www.rct.uk/collection/810077/procession-to-the-hustings-after-a-successful-canvass [29.10.2020] in which Fox's supporters carry both a placard of a fox and an effigy of a chicken.

17 See the Bronze medal (1796) [Anon.]; online: https://www.britishmuseum.org/collection/object/C_M-5054 [26.10.2020], and William Dent: *Election Compromise or a Cornish Hug in Westminster* (1790); online: https://www.britishmuseum.org/collection/object/P_J-4-29 [26.10.2020].

18 See Harriet Guest: Speech and Noise at the Westminster Elections. In: Republics of Letters 5 (2017), no. 2, p. 1–17. – See also Wiliam Dent: *The Man of Moderation* (1784); online: https://www.britishmuseum.org/collection/object/P_1868-0808-5149 [26.10.2020], and [Anon.]: *The ganders addressing the lion, to have the elephant remov'd* (1784); online: https://www.britishmuseum.org/collection/object/P_1868-0808-5161 [26.10.2020].

3 Animals and Carnivalesque Inversion

Politicians are still represented by animals today, as with the Democrats' donkey and Republicans' elephant in the US, which date from the nineteenth century. But these eighteenth-century texts imagining elections as populated by a menagerie of animals tapped into a contemporary understanding of elections as a time of carnivalesque inversion. Contemporary commentators and modern historians alike have characterised eighteenth-century elections as "a form of street theatre", a time of "public merriment, an almost carnival atmosphere" sometimes spilling over into small-scale disturbances or even "rioting of a quite deliberate nature". For the period of the election, normal social hierarchies could be suspended, when "the population cheered and jeered the candidates" who, although from the elite, 'might be openly ridiculed' by their social inferiors.[19] This was the 'world turned upside down', a very common feature of ballads, broadsides and prints across early modern Europe. Typically, these featured humans and nonhuman animals changing place: horses riding in carts being pulled by men, fish angling for people, rabbits chasing hunters, geese roasting cooks, and so on.[20] The use of animals in depictions of election scenes played on this association, giving a clear emphasis to the inversionary character of the electoral process. Some political prints even show successful candidates riding in carriages pulled not by horses, but by humans.[21] Another compounds the inversion by not only showing two of the candidates of the

19 Harry T. Dickinson: The Politics of the People in Eighteenth-Century Britain. Basingstoke 1995, p. 48. – See also O'Gorman, Campaign Rituals, p. 79–115; Frank O'Gorman: Ritual Aspects of Popular Politics in England (c. 1700–1830). In: Memoria y Civilización 3 (2000), p. 161–186.

20 Notably in an early nineteenth-century edition of *The World Turned Upside Down; or, No News, and Strange News* published in York by J. Kendrew, the title-page depicts *Tailor Riding a Goose*, an image derived, but thoroughly displaced, from Gillray's *Triumphal procession of little-Paul-the Taylor upon his new-goose*. See https://archive.org/stream/McGillLibrary-PN970_K4_W6_1820-1243/PN970_K4_W6_1820#page/n1/mode/2up [29.10.2020].

21 See Gillray's *Middlesex election (1804) – "A Long-Pull, a Strong-Pull, and a Pull-All-Together."* (1804) (note the dead cat in the foreground); online: https://www.britishmuseum.org/collection/object/P_1868-0808-13698 [26.10.2020] or Gillray's *The Entrance of Early Wolf (Lord Londsdale) into Blackhaven* (1792), featuring Londsdale as a wolf; online: https://www.npg.org.uk/collections/search/portrait/mw63422/James-Lowther-Earl-of-Lonsdale-The-pacific-entrance-of-Earl-Wolf-into-Blackhaven [29.10.2020]. – Uncoupling the horses from a favoured candidate's carriage and hauling it into the city was in fact a not uncommon electoral ritual: see for instance William Anthony Hay: Henry Brougham and the 1818 Westmorland Election: A Study in Provincial Opinion and the Opening of Constituency Politics. In: Albion: A Quarterly Journal Concerned with British Studies (2014), p. 38–39; online: https://www.jstor.org/stable/4054435 [26.11.2021].

1784 Westminster election as a fox (Fox) and a badger (Sir Cecil Wray), but having them being carried into battle by two women (the duchess of Devonshire and Mrs Hobart), while female ballad sellers look on and sing out their support for Fox.[22] In this print, the contest for who will take a seat in the nation's parliament is being fought only by females and quadrupeds: a complete inversion of the franchise (male human householders) but also a satirical commentary on the 'inadequacy' of both electors and candidates.

Historians since Mikhail Bakhtin have debated whether early modern carnivals that temporarily capsized hierarchies were 'safety valves' to allow social tensions to dissipate, or should be understood as more substantial challenges to power structures that could lead to real resistance and rebellion.[23] But the use of zoomorphism in electoral satire was the preserve neither of conservatives or radicals, and was adopted by both.[24] Candidates presented as dogs, geese, foxes, badgers or chameleons, or election clerks as monkeys, might each have been partisan, context-specific interventions in a particular election campaign. But they were also part of a broader discourse conceptualising the electoral process not as a routine and self-evidently constructive element of civil society, but as running counter to the normal order.

Metamorphosed candidates were far from the only electoral animals, and we can now look more closely at representations of the electoral 'crowd'. At contested elections, crowds congregated during the run-up to polling at locations connected with civic governance or with the candidates or their political parties, and then gathered at the 'hustings' where candidates could give speeches and voting took place. Contemporary depictions tend to show a diverse crowd, in terms of gender, age, socio-economic status and sometimes race, but notably, and in two of the images already discussed, they often include the presence of one or more dead cats. A famous instance is Robert Dighton's pen and watercolour drawing of the Westminster election of 1788, which features the

22 William Paulet Carey: *A meeting of the female canvassers in Covent Garden* (1784); online: https://www.britishmuseum.org/collection/object/P_1868-0808-5271 [29.10.2020].

23 E.g. Natalie Zemon Davis: Society and Culture in Early Modern France. Stanford 1965, ch. 4–6.

24 On 'radical' and 'conservative' uses and consequences of satirical representations of candidates, elections, and the electoral activity of 'the people'/particular groups (e.g. how such representations could work against bids to extend the franchise, and also play into different calls for 'reform'), see Packham, Literature and the Culture of Elections. – O'Gorman describes the conservative aspects of election rituals and ceremonies, and also how these rituals and ceremonies could be "agencies of political and social criticism" (O'Gorman, Campaign Rituals, p. 108).

body of a dead cat being thrown towards the hustings.[25] Other prints seem customarily to include the same motif: a dead cat being tossed around above an electoral crowd.[26] Perhaps giving an indication of what kind of person was responsible is another Robert Dighton image, this time of the 1796 Westminster election, in which a small boy is prominently placed at the foot of the hustings, holding a dead cat by a back leg, seemingly poised to launch it in the direction of the candidates.[27]

What should we make of this recurrent feature? Various sources suggest that dead cats were indeed thrown by particular members of actual electoral 'crowds'. In October 1710, Jonathan Swift described a menacing encounter with Whig supporters during the election for Westminster: "the rabble came about our coach, crying A *Colt*, a *Stanhope* [i. e. the names of the Whig candidates], &c. we were afraid of a dead cat, or our glasses broken, and so were always of their side".[28] In 1774, a report in the *Middlesex Journal* claimed that a dead cat "flung [...] amidst the spectators in Covent-Garden", "struck a Gentleman in the face, and beat out one of his eyes".[29] An often-repeated anecdote was told of Charles James Fox (seen in Dighton's 1796 image) who, on being told that a dead cat thrown at the hustings smelled worse than his namesake, a fox, replied "there was nothing extraordinary in that, considering it was a *poll*-cat".[30]

Evidence of this practice can also be found in the nineteenth century, for example, in different accounts of the 'chairing' ceremony (when the successful candidates were carried through the streets by their supporters) following the

---

25 See the detail from this in Corfield, Cats and Creativity, in the present volume (figure 12.4, p. 178).

26 E. g. *The Mask* and *The humours of Covent Garden or freedom of election*, both satirising the Westminster election of 1784; online: https://www.britishmuseum.org/collection/object/P_1868-0808-5226 [26.10.2020], and https://www.britishmuseum.org/collection/object/P_1868-0808-5232 [26.10.2020].

27 Robert Dighton: *The Westminster Election 1796*; online: https://www.britishmuseum.org/collection/object/P_1937-0508-37 [26.10.2020] (or for a detailed etching see https://www.britishmuseum.org/collection/object/P_1978-U-1907 [26.10.2020]).

28 Jonathan Swift: Letters, written by the late Jonathan Swift, D. D. Dean of St. Patrick's, Dublin, and Several of his Friends. From the Year 1710 to 1742. Vol. IV. London 1768, p. 34. – The account reflects Swift's own Tory sympathies (see n. 38 below). On Swift and the election, see George P. Mayhew: Swift's Political 'Conversion' and his 'Lost' Ballad on the Westminster Election of 1710. In: Bulletin of the John Rylands Library 53 (1971), p. 397–427.

29 Middlesex Journal, 18–20 October 1774, p. [2].

30 Francis Grose: The olio: being a collection of essays, dialogues, letters, Biographical Sketches, Anecdotes, Pieces of Poetry, Parodies, Bon Mots, Epigrams, epitaphs, & c. London 1792, p. 203.

victory of Viscount Sandon in the Liverpool by-election of October 1831.[31] One newspaper reported on the actions of a group of "ruffians", whose "object" was to "cause confusion, and then [...] plunder the persons of the unwary": "[t]hey had a dead cat, which they kept pitching among the crowd, to the huge delight of some, but to the great annoyance of the spectators generally".[32] Another paper, the *Liverpool Mercury*, suggested that a dead cat was thrown at (and struck) Sandon: "we regret to add that several personal insults were offered to his Lordship. [...] In Lord-street a stone was aimed at him; in Duke-street a piece of the bottom of a bottle was near hitting him; and soon after, we are assured (for we did not see it) a dead cat was flung with too accurate an aim".[33] The throwing of dead cats during public gatherings was not unique to elections, but it was evidently an established practice alongside and during election rituals and ceremonies, such as public meetings and 'chairing the member'.[34]

4 Animal Imagery and Its Meanings

What an individual was trying to achieve by throwing a dead cat at a candidate is of course difficult to ascertain. Famously, for Robert Darnton the slaughter of cats on the Rue Saint-Severin in 1730s Paris by apprentice printers was an act of revenge against their masters and perhaps an act of industrial protest.[35] For contemporaries, throwing dead cats during elections could certainly be seen as a form of political condemnation: the pro-reform *Liverpool Mercury* presented the reported attack on Sandon (who had objected to parts of the reform bill) as, at least partly, an expression of political "dislike".[36] What is clear is that the actual and imagined throwing of dead cats was strongly associated with the socially 'lowly' and election 'crowds'. The activity was, in effect, a metonym: a

31 On the by-election, see Margaret Escott: Liverpool; online: https://www.historyofparliamentonline.org/volume/1820-1832/constituencies/liverpool [26.10.2020].

32 Leicester Journal, 28 October 1831, p. [3].

33 Liverpool Mercury, 28 October 1831, p. 342.

34 For non-electoral examples, see e. g. Public Advertiser, 8 March 1775, p. [3]; Corfield, Cats and Creativity.

35 See Robert Darnton: The Great Cat Massacre and Other Episodes in French Cultural History. New York 1984. For a critique of Darnton, see Harold Mah: Suppressing the Text: The Metaphysics of Ethnographic History in Darnton's Great Cat Massacre. In: History Workshop 31 (1991), no. 1, p. 1–20.

36 Liverpool Mercury, 28 October 1831, p. 342; on chairings as occasions to express and encourage partisanship and electoral dissent, see Packham, Literature and the Culture of Elections. – On the *Liverpool Mercury*, see also Mark Harrison: Crowds and History. Mass Phenomena in English Towns, 1790–1835. Cambridge 1988, p. 227–230.

textual (or visual) figure that had come to stand for the electoral process. If an artist or author wanted a shorthand to represent an electoral crowd, the image of the dead cat was ready at hand. Thus, for instance, when Tobias Smollett has the hero of his picaresque novel *The Life and Adventures of Sir Launcelot Greaves* (1760–1761) encounter the supporters of rival candidates, one of the things that alert him to the venality and brutality of their county election is "a discharge or volley of stones, dirt, and dead cats; in consequence of which some teeth were demolished, and many surtouts defiled".[37] Those who threw carcasses might not be able themselves to vote, but their voice – for instance, at the hustings – had a political influence, and it was their public role that was being brought into question. References to and representations of dead cats could potentially signify the unfitness of 'the people' to engage in the electoral process, and still less to exercise democratic power.[38]

For those who wanted to deride the electoral participation of the 'humble', a key animal metaphor was the pig. It is true that prosperous voters, giving their support to whoever offered them most, could be satirised as pigs with their snouts in the trough.[39] But pig imagery was more often used to convey a strongly negative view of the electoral involvement of the socially 'lowly', activating traditional associations with, for example, gluttony, lust, ignorance, irrationality, and being easily led.[40] These ideas played into William Hogarth's influential election painting, *Chairing the Member* (1754–1755).[41] At centre-stage in Hogarth's image, a sow and four piglets hurtle downhill and, following a piglet

37 Tobias George Smollett: The Life and Adventures of Sir Launcelot Greaves. Ed. by David Evans. Oxford 1973, p. 72.

38 As Swift's account of the perceived violence of a specifically pro-Whig election 'rabble' suggests, such references and representations could be used to condemn particular electoral 'crowds' from a partisan standpoint; on the role of satirical representations (and other modes) in promoting different, positive and negative, views of the political agency of 'the people' (and particular people) during elections, see Packham, Literature and the Culture of Elections. On different attitudes towards 'crowds' in the late eighteenth and early nineteenth centuries, see Harrison, Crowds and History.

39 E.g. Gillray's *Political Amusements for Young Gentlemen: – or – The Old Brentford Shuttlecock, between Old Sarum, & the the Temple of St. Steevens* (1801), in which a print affixed to a wall depicts *The Old Sarum Electors* as a line of gorging pigs: https://www.britishmuseum.org/collection/object/P_1868-0808-6942 [29.10.2020].

40 On contemporary associations between swine and labouring-class individuals, see Spencer, Writing about Animals, ch. 6.

41 William Hogarth: *Chairing the Member* (1754–55); online: https://artuk.org/discover/artworks/an-election-4-chairing-the-member-123984 [29.10.2020]. – On Hogarth's paralleling of constituents with pigs, the symbolism of other animals, and portrayal of the 'crowd' in *Chairing the Member*, see Ronald Paulson: Hogarth. Vol. 3: Art and Politics, 1750–1764. Cambridge 1993, p. 163–167.

that has preceded them, are just about to plunge into a stream or sewer. Hogarth's pigs recall the Gadarene swine from the Synoptic Gospels of the Bible, destructive demons unleashed here not by Christ's exorcism but by the election. The pigs are a parallel to the human participants in the chairing, who are themselves careering towards their catastrophe: violence and disorder in the short term, and poor government and socio-economic ruin in the long. (The other nonhuman animals caught up in the chaos, such as a bear standing on hind legs, similarly serve to destabilise the barrier between human and nonhuman and so emphasise the 'animality' of the crowd. Even the goose, flying above the victorious candidate in parody of the traditional eagle, bears a striking resemblance to the candidate's bewigged head.)

Electoral swine imagery was given new life by the publication of Edmund Burke's *Reflections on the Revolution in France* in 1790. Burke warned that, among the consequences of the Revolution, "[a]long with its natural protectors and guardians [meaning the nobility and clergy], learning will be cast into the mire, and trodden down under the hoofs of *a swinish multitude*".[42] His pig metaphor, drawing on the Gadarene swine, immediately struck a chord with his conservative supporters and radical opponents, both of whom incorporated the image into their rhetoric as the Revolution crisis developed.[43] This imagery could be deployed as an insult levelled at people deemed unfit to play a role in electoral politics. The implicit reference was also to Christ's Sermon on the Mount: "Do not give what is holy to the dogs; nor cast your pearls before swine, lest they trample them under their feet, and turn and tear you in pieces."[44]

At the same time, however, Burke's phrase was also appropriated by political radicals as an ironic label for themselves and their supporters, as in Thomas Spence's *Pig's Meat, or Lessons for the Swinish Multitude* (published weekly, 1793–1795). Indeed, materials produced in response to elections up to (and beyond) the 1832 Reform Act reveal that Burke's swine imagery, and its wider reception, had a long-lasting impact on electoral culture. On the one hand it was deployed as a straightforward insult. As the essayist Vicesimus Knox observed, by 1795 (only five years after Burke's coining of the phrase) those

42 Edmund Burke: Reflections on the Revolution in France. London 1790, p. 117 (emphasis added).

43 For overviews of the complex and sustained engagement with Burke's swine imagery, see Don Herzog: Poisoning the Minds of the Lower Orders. Princeton 1998, ch. 12; Machell, Politics of Bestial Images, ch. 1; Spencer, Writing about Animals, ch. 6. Further useful commentary is in Olivia Smith: The Politics of Language. 1791–1819. Oxford 1984, p. 87–88; and Roland Bartel: Shelley and Burke's Swinish Multitude. In: Keats-Shelley Journal 18 (1969), p. 4–9.

44 Matthew, 7: 6.

in power "hardly condescend to recognize the idea of a People; but whenever they speak of the mass of the community, denominate them the mob, the rabble, or the swinish multitude", and this rhetoric was increasingly used by conservative candidates during and following the French Revolution crises.[45] On the other hand, addresses to 'the people' during particular electoral contests often reveal ironic engagement with, and explicit condemnation of, Burke's animal imagery. For example, the *Liverpool Mercury* reported a speech in support of radical candidate, Peter Crompton, made during polling for the 1820 Liverpool election by Egerton Smith, editor of the newspaper:

> Gentlemen, – You may suppose I am proud of being thus elevated above the "swinish multitude". But, I assure you, exalted as my present station is, it does not afford me such gratification as I experienced this day, when I had under one arm a poor dumb elector, and under the other a one-legged tailor, who both voted for the independent cause.[46]

In 1830, an electoral 'advice' letter in the *Liverpool Mercury* engaged with Burke to admonish the "labouring classes of Liverpool" to adopt behaviour in the upcoming mayoral and parliamentary elections that, it was claimed, would strengthen the case for extending the franchise.[47] The letter condemned Burke's "swinish multitude" as an "ungrateful" class slur. Yet it also reinforced charges of the 'animalistic' behaviour of 'the people', urging 'humble' constituents to "reform" their electoral conduct ("[y]ou can at any time get as beastly drunk as you like for half-a-crown; but act now like rational men") and thereby gain support for "the principle of universal suffrage".[48]

As the letter clearly indicates, Burke's animal imagery provided a kind of yardstick for measuring electoral behaviour, and was enlisted to expose contemporary electoral abuses. A report on the 1807 election for Surrey in the *Hereford Journal* took up Burke's controversial expression, claiming that "at the White Hart Inn, at Guildford, at the close of the Election", "[s]ome of the *swinish* multitude, not content with drinking as much as possible at the expense

45 Vicesimus Knox: The Spirit of Despotism. London 1795, p. 34–35.

46 Liverpool Mercury, 14 March 1820, p. 301. – Crompton was not returned; for details of the contest, see Escott, 'Liverpool'. On Egerton Smith, see Harrison, Crowds and History, p. 228–229.

47 Liverpool Mercury, 8 October 1830. – On advice pamphlets, see Knights, Representation, ch. 4; on the *Liverpool Mercury* and the wider Liverpool reform movement, with its emphasis on the "orderly conduct" of "the people", see Harrison, Crowds and History, p. 225–230.

48 See Liverpool Mercury, 8 October 1830, p. 326.

and liberality of Mr. Thornton, (one of the successful candidates) *pocketed* several bottles of wine".[49] By condemning the 'animality' of at least some constituents (whether franchised or unfranchised is unclear), the report, like Hogarth's *Chairing the Member*, potentially gave support to the view that 'humble' individuals should be excluded from the franchise.

Yet, as in Hogarth's election series, the actions of candidates were also held up to scrutiny: the report connected the 'swinish' behaviour of Surrey constituents with Samuel Thornton's electoral 'treating'. Indeed, Burke's animal imagery contributed to a long-established discourse within which the 'riotous' electoral behaviour of 'the people' was used to denounce the corrupt electoral influence of political elites, and the 'rotten' system that allowed such corruption to flourish.[50] An article printed in the *Hull Packet* attacked the influence of Earl Fitzwilliam in the Malton election of 1808, and painted a scathing portrait of the evils of a 'rotten' borough, asserting that "[w]hoever has attended to the habits, manners, disposition, and sentiments of a close Borough, must have remarked a lively illustration of [...] to use the [...] forcible language of the late Mr. Burke, a community of the 'swinish multitude' – a brutal ignorance, a sottish, sensual conduct, a gross servility, a disgusting obsequiousness".[51] The electoral 'vices' of 'the people' reflected those of Fitzwilliam in particular, and the mutually 'degrading' effects of corrupt electoral 'interference' in 'rotten' boroughs in general:

> a British Landlord, who, wishing still to exercise the authority of a Feudal Chief, to keep his tenantry in a state of vassalage, and to arm himself against whomsoever are in power, with the People in their right of the Elective Franchise, impresses on a close Borough its swinish character: [...]

49 Hereford Journal, 10 June 1807, p. [4].

50 On Andrew Marvell's engagement with and contribution to this discourse, see Kendra Packham: Marvell, Political Print, and Picturing the Catholic: An Account of the Growth of Popery and Arbitrary Government. In: Martin Dzelzainis and Edward Holberton (ed.): The Oxford Handbook of Andrew Marvell. Oxford 2019, p. 558–578, esp. p. 565–566, p. 568 (n. 60).

51 Hull Packet, 28 June 1808, p. [4]. Fitzwilliam was "sole patron" of Malton and "secured the return of his nominees unopposed until 1807, despite rumours of opposition at most elections" (Robert. G. Thorne: Malton; online: https://www.historyofparliamentonline.org/volume/1790-1820/constituencies/malton [29.10.2020]); see E. Anthony Smith: Earl Fitzwilliam and Malton. A Proprietary Borough in the Early Nineteenth Century. In: The English Historical Review (1965), p. 51–69). – Cf. the application of louse imagery to the Cornish pocket borough of Grampound in John Cartwright to Jeremy Bentham (18 February 1821) in: Stephen Conway (ed.): The Correspondence of Jeremy Bentham. July 1820 to December 1821. 12 Vols. Oxford 1994, Vol. 10, p. 287–288 (thanks to James Harris for this reference).

> by which he keeps human nature still in the mire, and by which he even degrades himself; for [...] he can only be regarded, at all events, as a sort of swineherd, if those whom he governs and directs, *he* converted into the degraded condition of a 'swinish multitude'.[52]

If the people were "swinish", this argument went, they were made so by the 'undue' influence of the powerful, and a corrupting – indeed 'dehumanising' – electoral system.

Inevitably, the corruption of the system was a common theme as the debate on political reform reached its climax. In June 1832, shortly after the passing of the Reform Act, an article appeared in *The Examiner*, entitled *Brutalities of the High and Low Compared.* The article again reported on "dead dogs and cats", this time thrown at the Duke of Wellington (who had fought so hard to block reform), when he "made a parade of his person in the City" six days earlier. Yet here, the violence of 'the people' was juxtaposed with the far greater violence of the state. When he had planned to suppress the crowd, the article explained, the Duke "did not think of putting dead cats and dead dogs into the hands of the soldiers, and ordering them [...] to be hurled at the discontented", but instead "doubtless had in contemplation charges with sharp bayonets, and the murderous volleys of round lumps of lead".[53] While it strengthened the association between throwing dead animals and the socially 'lowly', the article sharply contrasted this behaviour with the bloodshed committed by those of higher social rank. The paper condemned not a swinish multitude but a far more brutal, murderous elite.

5 Conclusion

Textual and visual representations of nonhuman animals had a significant discursive role in elections in long eighteenth-century England. Much of the zoomorphic satire focused on individuals, particularly election candidates. This feature is hardly surprising, since animal imagery has long been used to dehumanise opponents. The electoral crowd was also depicted in bestial terms, treating animals with cruelty, as well as being theriomorphic themselves (the 'swinish multitude'). But so too were the political elites who controlled or benefitted from elections. As the radical lawyer Thomas Erskine was said (by

52 Hull Packet, 28 June 1808, p. [4].

53 The Examiner, 24 June 1832, p. [401].

his opponents) to have put it in 1797, "if the People were a Swinish Multitude, those who represented them must necessarily be a Swinish representation".[54]

In fact, then, although it was deployed in a variety of ways, animal imagery often served to denigrate the electoral system as a whole. Elections were characterised as frenzied, inverted and dangerous occasions, when humans forgot their proper natures. These satirical animalisations were not straightforwardly radical or conservative (though they could, in specific contexts, be either). Nor were they necessarily arguing for electoral reform. Rather they were part of a sustained commentary on the electoral system that highlighted its eccentricity, irrationality and corruption. Such animal imagery reminds us that pre-Reform elections were not only, or straightforwardly, regarded with respect as an honoured part of the constitutional settlement, but could also be viewed as a wild, dehumanising, even bestial, moment of social and ontological disorder.

54 The Anti-Jacobin; or, Weekly Examiner, 4 December 1797, p. 30.

CHAPTER 12

"For I Will Consider My Cat Jeoffry": Cats and Creativity in Eighteenth-Century Britain

Penelope J. Corfield

Résumé

Les chats deviennent au cours du dix-huitième siècle des animaux domestiques très populaires, surtout dans les villes, alors que la Grande-Bretagne passe progressivement d'une société majoritairement agraire à une société plus urbaine. Pourtant, au cours de ces changements sociétaux, les chats ne perdent pas leurs pouvoirs magiques, comme le rappelle un célèbre folklore. Ils sont des amis familiers de l'homme, ronronnant au coin du feu, tout en gardant une forme d'étrangeté animale. Cette alliance de contraires est la source d'un grand élan de créativité littéraire et intellectuelle. Les chats semblent transmettre des messages d'aventure et de solidarité entre les espèces.

Abstract

Cats became very popular pets during the eighteenth century, especially in the cities, as Britain gradually moved from being a predominantly agrarian society to an increasingly urbanised world. Yet during these social changes, cats did not lose their magical powers, as many popular folklore tales bore witness. Cats, purring by the fireside, were familiar domestic friends, whilst retaining their relative feline aloofness and 'strangeness'. Their alliance of opposing characteristics was a source of great literary and intellectual creativity. Thus cats conveyed 'electric' messages both of adventure and of solidarity between humans and felines.

1 Introduction: The Cat Indoors

Cats are at once cosily familiar animals, yet remain aloofly independent.[1] They play, yet do not shed their dignity. As the mystic poet Christopher Smart (1722–1771) pertinently observed, in his tribute to his cat Jeoffry (c. 1759–1763): "For

1 See Sarah Brown: The Cat. A Natural History. Brighton 2020, passim.

 | DOI:10.1163/9789004495395_013

he is a mixture of gravity and waggery".[2] That feline conjunction of characteristics has over time sparked an immense amount of intellectual and literary creativity.

This analysis explores that theme in the context of the long eighteenth-century in Britain. The emphasis falls especially upon urban cats, in the context of the country's notable urbanisation – the most rapid in Europe, matched only in that era by the much smaller Dutch Republic.[3] So the plentiful farm cats and the rare Scottish wildcats do not feature centrally in the story,[4] although they confirmed the feline presence. The long eighteenth century was the era that celebrated the advent of domestic 'pets'. Cats were still expected to earn their keep, by catching mice. (Dogs also doubled as rat catchers). According to Daniel Defoe in 1722, every London house had at least one cat, while some had many more;[5] and the same applied elsewhere. But feline companionship was becoming newly prized over their traditional role as pest controllers.[6] The feline diet changed. Cats were no longer expected to live solely on captured mice. By the 1740s, special vendors toured town streets, selling meat scraps (usually horsemeat from the knackers' yards) for pets. That familiar sight eventually disappeared in the 1920s, but only when tinned cat food took over the market.[7]

Needless to say, dogs also remained immensely popular throughout the eighteenth century in Britain. They are highly versatile pack animals. They undertook a variety of companionate roles, from leading the outdoors hunt

2 Christopher Smart: Jubilate Agno [c. 1759–63]. In: David Wheeler (ed.): A Selection of Poetry. Raleigh/NC 2011, p. 43 and p. 123.

3 See Penelope J. Corfield: The Impact of English Towns. 1700–1800. Oxford 1982. p. 4–9; Jan de Vries: European Urbanisation. 1500–1800. London 1982, p. 44–48 and p. 258–260.

4 See Morris Allan and Tim Tyler: In Search of the Scottish Wildcat. Ed. by Lillian King. Kelty 2002, p. 1–120.

5 Daniel Defoe (1660–1731): A Journal of the Plague Year [1722]. Ed. by Cynthia Wall. London 2003, p. 118.

6 For more information, see variously Keith Thomas: Man and the Natural World: Changing Attitudes in England. 1500–1800. London 1983; Rob Boddice: A History of Attitudes and Behaviours towards Animals in Eighteenth- and Nineteenth-Century Britain. Anthropocentrism and the Emergence of Animals. Lewiston 2008; Ingrid H. Tague: Animal Companions. Pets and Social Change in Eighteenth-Century Britain. Philadelphia 2013 (Animalibus Vol. 5).

7 A case at London's Old Bailey in 1748 cited: "a woman that sells cats and dogs meat". – Cf. Old Bailey Proceedings Online; trial of John Walker, April 1748; online: https://www.oldbaileyonline.org/browse.jsp?id=t17480420-1&div=t17480420-1#highlight [10.12.2020]. – In 1861, Henry Mayhew estimated that London contained c. 1,000 cats-meat vendors (see Sarah Hartwell: 'The Cat's Meat Man'; online: http://messybeast.com/cats-meat-man.htm [10.12.2020]).

FIGURE 12.1 Angora Cat, very much at home in a sumptuous domestic interior. From G.-L. Leclerc, comte de Buffon and L. J. Daubenton, *Histoire naturelle* (Paris 1756), Vol. 6, plate 5.

for foxes to sitting indoors as ladies' lap-dogs.[8] They were repeatedly said to be 'man's best friend'.

Yet the popularity of dogs did not preclude the parallel success of cats. The feline tribe did not join in human sports or walks. And there was no concept of an ornamental 'lap-cat', even though cats often sat in laps. The feline mystique was to retain the appeal of the domestic familiar, whilst still appearing strange, autonomous, independent. In this period, the black cat in particular was losing its traditional malign identification with witchcraft, at least in educated circles. But the sable felines retained an association with luck – whether good or bad luck depended on the circumstances – and thus they were not creatures to be crossed. So cats made the transition from being utilitarian mousers to domestic familiars, purring by the eighteenth-century's newly ubiquitous coal-fires, whilst still retaining an aura of magic.[9]

One key association was particularly notable. Cats were and are excellent companions for writers and thinkers. The feline capacity for patient watching and waiting was a great asset, during the long hours of composition. And cats did not clamour to be taken for walks. Instead, as shown in further detail below, they were (and remain) not only fine companions to literary and intellectual endeavour, but also magnificent prompts to creativity.

The real-life eighteenth-century cat with the greatest claim to posthumous fame was undoubtedly the black cat Hodge belonging to Dr Johnson, the compiler of the celebrated *Dictionary of the English Language* (1755). This animal was not in fact the only feline pet in the household. But he was considered to be Johnson's favourite. And his death was mourned in a special *Elegy* (written c. 1764; published 1778), by a family friend, Percival Stockdale. He praised the cat's simple and blameless life, not rushing after changing fashions but wearing the same black coat throughout. And Stockdale wryly urged his fellow humans:

> Let virtue in thy bosom lodge;
> Or wish thou hadst been born a Hodge.[10]

Clearly, there was no hint of demonic witchcraft here.

8 See Brian Hare and Vanessa Woods: The Genius of Dogs. Discovering the Unique Intelligence of Man's Best Friend. New York 2013, passim.

9 For further information, see variously Mary Oldfield Howey: The Cat in Magic and Myth [1930]. Mineola NY 2003; Martha Gray: Grimalkyn, the Witch's Cat. Power Animals in Traditional Magic. Lanham 2013.

10 Percival Stockdale: An Elegy on the Death of Dr Johnson's Favourite Cat [c. 1764; 1st ED 1778]. Lichfield 1997, p. 1–4.

Since then, Hodge has had a lively cultural after-life, featuring in stories, illustrations, and a dramatic monologue.[11] And in 1997 his image resumed his watching, in the form of a statue, located in Gough Square, outside the central London townhouse which Dr Johnson rented between 1748 and 1759. Hodge sits on a replica of Johnson's *Dictionary of the English Language* (1755), next to the shells of oysters, with which he was fed by Dr Johnson, as reported by the lexicographer's friend and most famous biographer James Boswell. Sometimes tourists place coins on the plinth or hang ribbons on the statue, for good luck. So this black cat's connotations are now those of through-time benevolence.

2 The Familiarity of Cats

"A blazing fire, a warm rug, the candles lit and curtains drawn, the kettle on for tea, and finally the cat before you, attracting your attention – it is a scene which everybody likes, unless he has a morbid aversion to cats, which is not common".[12] Thus in 1834 the writer Leigh Hunt envisioned a classic domestic interior. In another essay, he confessed that he himself was not a great felinophile. Yet that was not relevant to his general point. Cats by the fireside were prime signifiers of household contentment. They combined traditional magic with a new purring domesticity. Hence, unsurprisingly, cats featured in many children's rhymes, some new minted in this period, like *The Cat & the Fiddle* (first found in print in 1765)[13] – later recycled into a popular name for an inn or tavern.[14]

Various eighteenth-century poets were also inspired to write verses to the feline tribes, large and small. Their ranks included Anne Finch; John Gay, James Thomson; Thomas Gray; Christopher Smart; Percival Stockdale; Anna Seward; William Cowper; William Blake; William Wordsworth; P. B. Shelley; Joanna Baillie; John Keats; and John Clare. (A smaller number wrote poems to dogs too: including Alexander Pope's whimsical epigram engraved on the collar of a puppy presented to the Prince of Wales, c. 1737).[15] Such expressivity about

11 For affectionate examples, see Yvonne Skargon: Lily and Hodge and Dr. Johnson. Truro 1991; Kirsten Tambling: Hodge's History of Cats. Cirencester 2014.

12 Leigh Hunt: The Cat by the Fire [1834]. In: idem: Essays. London 1883, p. 263.

13 Jeremy Barlow: The Cat and the Fiddle. Images of Musical Humour from the Middle Ages to Modern Times. Oxford 2006, passim.

14 The Cat & Fiddle Inn, in the Cheshire/Derbyshire Peak District (built 1813), long flourished as England's second highest public house (closed 2015).

15 See John Butt (ed.): The Poems of Alexander Pope. London 1965, p. 826.

'dumb animals' was part of the newly cultivated sensitivity to the authentic concerns of other living beings.[16]

Authors cogitating at their desks had good opportunities to study their companionate cats. Feline behaviour was thus closely observed. Christopher Smart, who lived in a world where human bodily cleanliness was often a struggle to achieve,[17] particularly appreciated his cat Jeoffry's careful daily toilet: "For he is the cleanest in the use of his forepaws of any quadruped".[18]

Others enjoyed the feline mix of snoozing calm alternating with quick playfulness. Both were admired qualities. Anna Seward's poem imagining *An Old Cat's Dying Soliloquy* (1792) began with fond reminiscences of life with a loved master, termed Acasto in mythic vein:

> Years saw me still Acasto's mansion grace,
> The gentlest, fondest of the tabby race;
> Before him frisking thro' the garden glade,
> Or at his feet, in quiet slumber laid [...].[19]

And in 1801 the observant William Wordsworth caught the scampering intensity of a young cat, trying to catch the falling leaves of autumn:

> But the Kitten, how she starts,
> Crouches, stretches, paws, and darts!
> First at one, and then its fellow,
> Just as light and just as yellow.
> There are many now – now one,
> Now they stop and there are none:
> What intenseness of desire,
> In her upward eye of fire![20]

16 See James Serpell: In the Company of Animals. A Study of Human-Animal Relationships. Oxford 1986; Ingrid H. Tague: Dead Pets: Satire and Sentiment in British Elegies and Epitaphs for Animals. In: Eighteenth-Century Studies 43 (2008), no. 3, p. 289–306; Arnold Arluke and Clinton Sanders: Between the Species. Readings in Human-Animal Relationships. Boston 2009.

17 See Peter Ward: The Clean Body. A Modern History. Montreal 2019, esp. p. 18–29 and p. 107–112.

18 Smart, Jubilate Agno, p. 43 and p. 123.

19 Anna Seward: The Old Cat's Dying Soliloquy [1792]. In: The Poetical Register and Repository of Fugitive Poetry, for 1803. 2nd ed. London 1805, p. 425.

20 William Wordsworth: The Kitten and the Falling Leaves [1801; publ. 1807]. In: Thomas Hutchinson (ed.): The Poetical Works of William Wordsworth. Oxford 1926, p. 170–171.

This poem was not outstanding among Wordsworth's massive output. But his affectionate empathy with the animal's physical absorption was well conveyed; and it was one that readers could readily share.

Notable feline characteristics also included the cats' capacity for happy purring; their love of milk and cream; their velvet paws which can unleash sharp claws; and their twitching tails which convey their changing moods. As small and generally amenable domestic familiars, wrapped in fur, they made fine pets for children. "O Cat beloved!/ Thou dear companion of my tender years!" ran one poem, imagining the distress of a young girl, when being sent to boarding-school and thus parted from her pet.

> How shall I eat while you are not beside
> To share the bit? How shall I ever sleep
> While I no more your lulling murmurs hear?[21]

This verse, by the Scottish bard James Thomson, clearly was expected to be received sympathetically.

Elderly and single people, especially »the widowed dame or lonely maid«, as Joanna Baillie specified in a poem of 1810,[22] were equally likely to be cat lovers. Eighteenth-century artists and satirists also acknowledged the female interest in felinity, which was assumed to be a characteristic of the sex. Thus countless women and young girls were shown as cuddling cats or kittens.[23] It was an established visual trope for feminine amiability and charm, with perhaps a hint of vulnerability.

Plenty of men, however, owned and appreciated cats. Well-known cases included not only the already-cited Dr Johnson and Christopher Smart but also the writer and art connoisseur Horace Walpole;[24] the legal philosopher Jeremy Bentham – who was one of the first protagonists of animal rights;[25] the

---

21 James Thomson: Lisy's Parting with her Cat [ca. 1716]. In: James Logie Robertson (ed.): The Complete Poetical Works of James Thomson. Oxford 1908, p. 512. – The girl was probably Thomson's sister Elizabeth.

22 Joanna Baillie: The Kitten [1810]. In: eadem: Fugitive Verses. London 1840, p. 197.

23 See Amy Freund and Michael Yonan: Cats: The Soft Underbelly of the Enlightenment. In: Journal 18: A Journal of Eighteenth-Century Art and Culture 7 (2019); online: https://www.journal18.org/issue7/cats-the-soft-underbelly-of-the-enlightenment/ [10.12.2020]; Caroline Bugler: The Cat. 3500 Years of the Cat in Art. London 2011.

24 For Horace Walpole cf. Christopher Frayling: Horace Walpole's Cat. London 2009.

25 For the cat named Sir John Langborn, belonging to Jeremy Bentham, cf. Adam White: Heads and Tales. Or, Anecdotes and Stories of Quadrupeds and Other Beasts [1870]. Whitefish/MT 2008, p. 150–151.

FIGURE 12.2 Thomas Rowlandson's satirical print of A Catamaran (his own verbal joke): An Old Maid's Nursery (1803) shows a venerable dame, festooned with a dozen attentive cats and sundry kittens, whilst a nursemaid spoon-feeds yet another cat which is swaddled like a baby.

Poet Laureate Robert Southey, whose home at Greta Hall in Keswick was full of cats;[26] and the novelist Sir Walter Scott, whose tabby named Hinse (sometimes spelled as Hinx) reportedly tyrannised over his pack of dogs.[27] An accumulating legend further identifies Sir Isaac Newton as not only a cat-lover but as the inventor of the cat-flap, allowing cats to enter and leave their domiciles at will. However, while that story agreeably humanises the great physicist, there is no evidence that he actually kept a pet.[28] Interestingly, there are not many eighteenth-century visual images of these men with their feline companions. Perhaps the imagery was seen as either too 'female'; or, if male, then too 'wizard-like'.

But, as popular stories about a hero from earlier times recounted, a feline ally was far from unmanly. A picaresque tale celebrated the famous Dick Whittington and his cat, whose prowess as a rodent controller helped to make Whittington's fortune.[29] The message was meritocratic. Loosely based upon a real-life Whittington from medieval times, the story celebrates a friendless poor boy who rose, through persistence and ingenuity, to become Lord Mayor of London. His trusty feline friend stands by him throughout. By the eighteenth century, their joint adventures were recycled in improving tales, children's shows and popular pantomimes. Indeed, so compelling and reassuring was the tale that variants have appeared in many cultures, from early times to this day.[30]

26 Robert Southey: Memoir of the Cats of Greta Hall [Keswick]. In: idem: Select Prose of Robert Southey. Ed. by Jacob Zeitlin. New York 1916, p. 305–309.

27 Walter Scott was depicted at his writing desk, with Hinse sleeping by him, and a dog at his feet: Sir Walter Scott in his Study at Castle Street [Edinburgh], by John Watson Gordon (c. 1840s) in Scottish National Portrait Gallery.

28 This legend about Isaac Newton dates from the nineteenth century. Newton was said to have made one large hole in his Cambridge study door for the mother cat and three small ones for her three kittens, allowing later generations to laugh at the great man's alleged lack of common sense.

29 For examples, see [Anon.]: The Famous Story of Dick Whittington and his Cat: Showing how from a Poor Country Boy, he Attained Great Riches, and [became] … Lord Mayor of London. London 1788; George Cruikshank, The History of Dick Whittington … and the Adventures of his Cat (c. 1820).

30 See Thomas Keightley: Tales and Popular Fictions. Their Resemblance and Transmission from Country to Country. London 1834, p. 241–266.

3 The Strangeness of Cats

Single-mindedness of intent was (and remains) a famous feline quality. The cat's autonomous abilities, which are not readily susceptible to human direction, give them a reputation for independence, sometimes interpreted as guile. There were (and are) no feline equivalents of trained guide dogs, hunting dogs, tracker dogs, sheep-dogs, police detection dogs, guard dogs, or rescue dogs. Cats go their own way. And often (not invariably) get their own way. Hence the phrase 'herding cats' is used to indicate the near-impossibility of getting a group of high-spirited, independent-minded individuals to work as a team.

That feline reputation for wily self-interest was endorsed in a 1713 poem by Anne Finch, Countess of Winchilsea. Entitled *The Eagle, the Sow and the Cat*, the verses outline the manoeuvres between this unlikely trio of animals: an eagle nested at the top of an oak tree, the sow with her piglets, snug among the roots, and the cunning puss in between.[31] She, silkily, stirs trouble between them. The sow and eagle panic and flee. As a result, the cat reigns over the tree, complete with tasty eagle chicks and piglets. Finch's moral is that people should beware smooth-talking, outwardly friendly advisers, before it is too late.

Another component of the strangeness of cats comes from association with their majestic relatives. Lions and tigers are grander, wilder, and more ferocious than the fireside tabbies. But the ancestral relationship gives cachet to the domesticated junior cousins.[32] Among Jeoffry's good points, Christopher Smart commented: "For he is of the tribe of Tiger".[33] Joanna Baillie also noted feline kinship. The kitten's glaring eyes and rapid pounce recalled the looks and gait of ruthless lions and tigers.[34] All cats had the same unblinking gaze. Hence William Blake, when debating the origins of *The Tyger* (1794), asks rhetorically: "In what distant deeps or skies/ Burnt the fire of thine eyes?"[35]

Little wonder that cats had a long association with magic, for both good and ill. Their proverbial nine lives were also evidence of a reputation for uncanny behaviour. They were agile and wily enough to escape death repeatedly.

---

31 Anne Kingsmill Finch, Countess of Winchilsea: The Eagle, the Sow and the Cat. In: eadem: Miscellany Poems on Several Occasions. London 1713, p. 212–215.

32 See Alan Turner: Big Cats and their Fossil Relatives. An Illustrated Guide to their Evolution and Natural History. New York 1997, passim.

33 Smart, Jubilate Agno, p. 42 and p. 122.

34 See Baillie, The Kitten, p. 199.

35 William Blake: The Tyger [1794]. In: Kathleen Raine (ed.): A Choice of Blake's Verse. London 1970, p. 61.

FIGURE 12.3 Detail showing pet cat, its teeth bared and claws flexed, gazing avidly at a caged bird, from William Hogarth's portrait of *The Graham Children* (1742)

A 'grimalkin' (originally meaning a grey cat) was the reputed assistant to a witch or sorcerer. Their power was potentially malign.

On the other hand, cats in folk tradition were often helpful, and were later translated into amiable pantomime characters. The Italian 'Gatto con gli stivali' became in French 'Le Maître Chat ou Le chat botté', in German 'der gestiefelte Kater'; and in English 'Puss in Boots'. Aware of the cats' strange abilities, Joanna Baillie accordingly wondered: "Whence hast thou then, thou witless puss!/ The magic power to charm us thus?"[36] She suggested that the kitten's playfulness was analogous to the appeal of a sportive child, remembered in people's hearts long after cats had grown up to face "full many a cuff and angry word" if they misbehaved.

As that sharp comment suggested, real-life pets were not always pampered. Nor were poetic reactions homogeneous. Some authors empathise with their cats. Others were relatively sour. John Keats stressed, in his poem *To Mrs*

36 Baillie, The Kitten, p. 198.

Reynolds Cat (1818), that the pretty puss with its soft fur was in fact a hardened killer:

> Cat! who hast passed thy grand climacteric,
> How many mice and rats hast in thy days
> Destroyed? How many tit-bits stolen? Gaze
> With those bright languid segments green, and prick
> Those velvet ears – but prithee do not stick
> Thy latent talons in me [...].[37]

In effect, the poet warns readers not to be deceived by dulcet exteriors. The cat's habit of toying with its prey was another trait which was deplored by sensitive humans, even by those who were admirers of the species. But Christopher Smart found at least a partial excuse for Jeoffry: "For when he takes his prey he plays with it to give it a chance./ For one mouse in seven escapes by his dallying".[38]

Meanwhile, Shelley also used a feline metaphor to point a social moral for all living beings. His epigrammatic *Verses on a Cat* (1800) – one of his juvenile efforts, saved for posterity by his sister – stressed that causes of suffering are variegated:

> You would not easily guess
> All the modes of distress
> Which torture the tenants of earth.[39]

In one case, the solution was clear. The distressed cat wanted a rat. But resolving the range of human desires was tricky:

> And which is the best
> I leave to be guessed [...].[40]

37 John Keats: To Mrs Reynolds Cat [1818]. In: William Garrod Heathcote (ed.): The Poetical Works of John Keats. Oxford 1956, p. 534.

38 Smart, Jubilate Agno, p. 42 and p. 122.

39 Percy Bysshe Shelley: Verses on a Cat [1800; publ. 1858]. In: Thomas Hutchinson (ed.): The Chief Poetical Works of Percy Bysshe Shelley. London 1907, p. 829.

40 Ibid.

This word-play, intended in fun, showed how domestic felinity prompted youthful meditations.

So great were the numbers of real-life cats in eighteenth-century Britain that their experiences were diverse. While strays prowled the streets and alleys, beloved pets were treated well. One sign of that diversity was the occasional sight of dead cats and dogs in the streets. All towns, large and small, were full of animals;[41] and they did not all die in abattoirs. So the trope of dead cats lying in gutters was an eighteenth-century poetic signifier of urban degeneration.[42] In fact, it was rare for town visitors and residents to mention the visibility of animal corpses, partly through familiarity with the sight – and partly because cadavers were eventually removed by foraging rats, foxes and town scavengers.

Nevertheless, at times eighteenth-century street politics were rendered more aggressively hostile by the throwing of dead cats and dogs at opponents. Generally, hostile crowds used as projectiles whatever urban detritus came to hand, such as stones and oyster shells. But, when available, dead cats and dogs were also thrown. Targets included political rivals during election affrays, or malefactors in the pillory who were unpopular with the witnessing crowds. Such actions were sufficiently part of an urban scuffle for a lawyer in 1787 to ask a witness: "It is a part of the fun of the day to push and shove, and throw dead cats, and make it [the movement of the crowd] as inconvenient as possible?" to which the answer was simply: "Yes".[43] Moreover, the shock value of disrespecting animal corpses was growing with the rising status of pets.[44] A dead cat flying through the air symbolised bad magic – and an unruly crowd.

41 For further contextual information, see Thomas Almeroth-Williams: City of Beasts. How Animals Shaped Georgian London. Manchester 2019.

42 See Jonathan Swift: Description of a City Shower [1710]. In: Herbert Davis (ed.): Swift. Poetical Works. Oxford 1967, p. 93; and context in Penelope J. Corfield: Walking the City Streets. The Urban Odyssey in Eighteenth-Century England. In: Journal of Urban History 16 (1990), no. 2, p. 132–174.

43 Old Bailey Proceedings Online, 12th December 1787, trial of John M'Carty and Thomas Hartman; online: https://www.oldbaileyonline.org/print.jsp?div=17871212 [10.12.2020].

44 A case in 1730s Paris caused outrage, when some irate apprentices killed their mistress's cats, having first symbolically tried them for witchcraft. – Cf. Robert Darnton: The Great Cat Massacre and Other Episodes in French Cultural History. New York 1984, p. 75–106; and critique in Harold Mah: Suppressing the Text. The Metaphysics of Ethnographic History in Darnton's Great Catt Massacre. In: History Workshop 31 (1991), no. 1, p. 1–20.

FIGURE 12.4 Robert Dighton's *Westminster Election* (1788) depicts the characteristic cross-class election hurly-burly, including a youngster trying to pick a pocket (Centre), whilst a dead cat has been thrown (upper L) as a signal mark of crowd disrespect and irreverence

4 Cats and the Literary Imagination

Their mixture of domestic familiarity and characterful strangeness made cats into a potent spur for literary creativity. Of course, the linkages cannot be proved with the accuracy required in (say) a court of law. Authors rarely explained their creative processes. One of the most specific references to the stimulus provided by an attentive cat came from Christopher Smart, when he mused of Jeoffry: "For he is good to think on, if a man would express himself neatly".[45] Such a comment, while suggestive, falls well short of proof. And authors were and remain a heterogeneous group. One apparent eighteenth-century literary link, by the way, had nothing to do with real felines. The Kit-Cat Club, popular among leading Whig authors in the 1710s, was so named not after the cavorting of cats but after the owner of inn where they initially met. He was Christopher (nicknamed Kit) Catt, vendor of famous mutton-pies.[46]

It is, however, possible to make some inferences about the developing literary tradition of writing about cats. Stories and songs abounded. And, in particular, there was a notable tendency to anthropomorphise feline characters who straddle the species boundary. They featured among the many talking animals which were imagined in the eighteenth century, appearing not only within the new genre of children's literature but among adult writings too.[47]

Joanna Baillie, herself a felinophile who moved in literary circles, defined one characteristic category of cat owner as:

> The learned sage, whose thoughts explore
> The widest range of human lore,
> Or with unfetter'd fancy fly
> Through airy heights of poesy [...].[48]

The moralist William Cowper (1731–1800) agreed. His poem to *The Retired Cat* (written 1791) saluted an ideal literary companion: "A Poet's Cat, sedate and

45 Smart, Jubilate Agno, p. 42 and p. 124.

46 For context, see Ophelia Field: The Kit-Cat Club. Friends who Imagined a Nation. New York 2008.

47 For general context, see variously Andrew O'Malley: The Making of the Modern Child. Children's Literature and Childhood in the Late Eighteenth Century. New York 2003; Matthew O. Grenby: Children's Literature. Edinburgh 2008; and Tess Cosslett: Talking Animals in British Children's Fiction, 1786–1914. Aldershot 2006.

48 Baillie, The Kitten, p. 197.

grave,/ As Poet well could wish to have". It transpired that the puss, like her owner, liked to philosophise – in her case, with her eyes shut. She thus climbed into a half-open drawer to snooze, only to become imprisoned when the tidying chambermaid, unawares, closed the drawer. Eventually the hungry, disconsolate cat is released. But she has learned the invaluable lesson that the world does not revolve around her. So Cowper drew a pointed moral: "Beware of too sublime a sense/ Of your own worth and consequence!"[49]

Of course, feline-human linkages were far from universally positive. Looking back to earlier times, William Shakespeare referred frequently to cats, his tone ranging from the pejorative to the mundane. In *Hamlet*, for instance, the hero notes that, even amid disasters, ordinary life continues: "The cat will mew and [every] dog will have his day".[50] The feline soundtrack signalled normalcy.

However, some writers began to experiment with alternative viewpoints. One printer's assistant, William Baldwin, created a pioneering fiction which included talking cats, derived from folk tales. For him, intelligent felines could understand humans and, dangerously, read their secrets. Hence Baldwin's title was *Beware the Cat*, (written 1553), as he warned that people should behave prudently, since they lived before ever-present witnesses.[51] His tale had only a muffled impact. But it survived as a curiosity, being reprinted in 1786 and again in 1860.[52] And in France, Baldwin's contemporary, the philosophic essayist Michel de Montaigne, also tried to step outside his species carapace. "When I play with my cat, how do I know that she is not passing time with me rather than I with her?"[53] What did the manifest feline consciousness actually portend? It was a relevant query in Montaigne's quest to deflate human species-arrogance.

49 William Cowper: The Retired Cat [1791]. In: William Hayley (ed.): The Life and Posthumous Writings of William Cowper, Esq. With an Introductory Letter to the Right Honourable Earl Cowper. Vol. 1. New York 1803, p. 253 and p. 258.

50 William Shakespeare: Hamlet. Prince of Denmark [c. 1599–1600], Act 5, sc. 1. In: Bretislav Hodek (ed.): William Shakespeare. The Complete Works. Comprising his Plays and Poems. London 1970, p. 977.

51 See William Baldwin: A Marvellous History. Entitled Beware the Cat: Containing Divers Wonderful and Incredible Matter […] [1570]. In: William A. Ringler (ed.): Beware the Cat: The First English Novel. San Marino/CA 1988.

52 First reprinted in Joseph Ames and William Herbert: Typographical Antiquities: Or, an Historical Account of the Origin and Progress of Printing in Great Britain and Ireland. London 1786.

53 Michel de Montaigne: Essais [1580; transl. into English 1613]. In: Michel de Montaigne: The Complete Essays, transl. and ed. Michael Andrew Screech. London 1987, book 2, essay 12, p. 505.

Gradually, cats were moving from folk tales and nursery rhymes into the imaginative landscape of fiction. Daniel Defoe's *Robinson Crusoe* (1719) had two cats with him, when he escaped onto his desert island. But the castaway's attitude remained signally prosaic. When his pets interbred with wild cats and produced too many young, Crusoe shot some and banished the rest from his house. He kept only two favoured in-house companions, whose subsequent offspring he ruthlessly shot or expelled from his little domain.[54] Such practicality was far from romance. Yet his cats, dog and parrot provided companionship, which Crusoe appreciated.

Numerous poets, however, not only observed their feline friends but were among the first to re-imagine the world from the feline viewpoint. So Thomson's account of the schoolgirl's woe at leaving her pet was matched by the cat's lament at losing a prime supplier of titbits, cream, and kindly stroking: "Beneath whose soothing hand/ Soft shall I purr?" A life without love would lack all meaning: "What is the dull officious world to me?"[55]

Similarly, Seward's *Old Cat's Soliloquy* stressed the grief of prospective loss. The tabby mourned the onset of old age ("pain has stiffened these once supple limbs"); but chiefly sorrowed to leave her benevolent master: "Ne'er shall thy now expiring puss forget/ To thy kind care her long-enduring debt".[56]

Such expressions of tenderness provided a positive affirmation of cat-human bonds. And even when the imagined feline consciousness was primarily focused upon catching mice, it was still granted its own validity. So John Gay's humorous poem on *The Rat-Catcher and Cats* (1727) – one of his many animal fables – mused on the professional rivalry to keep Britain's larders free of rodents. Without any feline competitors, the human rat-catchers would become "Sole guardians of a nation's cheese!" They could then increase their charges. However, a resident cat tries to foil his human rival. A confrontation ensures. The cat is about to be killed, when a wise feline brokers a truce. The rivals agree to split the work, on the grounds that: "There's game enough for us and you".[57]

Getting inside the mind of a cat was an exercise in trans-species empathy. One short tale in 1760 explored *The Life and Adventures of a Cat*. Its stance was somewhat facetious but genial. The hero was a handsome Tom Cat. His

54 See Daniel Defoe: The Life and Strange Surprising Adventures of Robinson Crusoe [...]. Written by Himself [1719]. Ed. by John Richetti. London 2001, p. 61, p. 99, and p. 144.

55 Thomson, Lisy's Parting with her Cat, p. 512.

56 Seward, Old Cat's Soliloquy, p. 425–426.

57 John Gay: The Rat-Catcher and Cats [1727], Fable 27. In: Vinton A. Dearing (ed.): John Gay: Poetry and Prose. Vol. 2. Oxford 1974, p. 330–331.

ancestry was surely as good as "the famous Cat of Montaigne, who laughed at her Master, for being such a fool, as to spend his time in playing with her, and actually told him so". This Tom has a message too. He inveighs against the human invention of the mouse-trap: "It is a monopoly, contrary to the nature and reason of things, to substitute an inanimate piece of wood instead of a living creature, made and intended for the purpose".[58]

Collectively, writings and imaginings of this ilk were contributing to the slow but long-term cultural shift, which ceased to assume that humans had the right to dominate and wantonly mistreat all 'inferior' species. In Britain attitudes were changing – especially in middle-class circles – well before the formation in 1809 of the first Society for Preventing Wanton Cruelty to Brute Animals.[59] In the 1780s, for example, Jeremy Bentham urged that humans had a responsibility to all sentient beings: "The question is not: Can they reason? nor, Can they talk? but, Can they suffer?" And he speculated further that: "The day may come, when the rest of the animal creation may acquire those rights which never could have been withholden from them, but by the hand of tyranny".[60]

To repeat: it cannot be said that all instances of literary and intellectual creativity were cat-inspired. Such a proposition goes far too far. Jane Austen immediately provides a case of cool indifference. Her sole fictional reference to cats occurs in *Sense and Sensibility* (1811), where the effusive Mrs Jennings sighs that, without company, "Lord! we shall sit and gape at one another as dull as two cats".[61] (This remark was sly passing jest from Austen, since her fictional Mrs Jennings is always chatty and real-life cats can be very vocal).

Nonetheless, a distinctive and accumulating minority strand in literary consciousness was detectable.[62] The fictional cat was both talking and acquiring engaging characteristics which were human-related but not-quite-human.

58 [Anon.] [attrib. Henry Fielding]: The Life and Adventures of a Cat. London 1760, p. 15 and p. 185–186.

59 See Harold D. Guither: Animal Rights. History and Scope of a Radical Social Movement. Carbondale/IL 1998; Aaron Garrett (ed.): Animal Rights and Souls in the Eighteenth Century. Bristol 2000; and Diana Donald: Women against Cruelty. Protection of Animals in Nineteenth-Century Britain. Manchester 2019.

60 From 1823 annotated edition of Jeremy Bentham: An Introduction to the Principles of Morals and Legislation [1789]. London 1823, ch. 17, sect. 1, n. 122. – See also Johannes Kniess: Bentham on Animal Welfare. In: British Journal for the History of Philosophy 27 (2019), no. 3, p. 556–572.

61 Jane Austen: Sense and Sensibility [1811]. Ed. by Ros Ballaster. London 2003, p. 262.

62 For further context, see Howard Loxton: 99 Lives. Cats in History, Legend and Literature. London 1999.

Later examples include Edward Lear's nonsense poem *The Owl and the Pussycat* (1871);[63] and, just before that, Lewis Carroll's enigmatic Cheshire Cat in *Alice's Adventures in Wonderland* (1865).[64] That animal's mythic quality of appearing and slowly disappearing as it pleased confirmed its cultural magic, which remains much commemorated.[65] And it was the philosophic forerunner of the physicist Erwin Schrödinger's equally ambiguous quantum cat, first postulated in 1935. That mysterious animal also has a long intellectual afterlife.[66] The feline conjunction of familiarity plus strangeness is thus continually being updated from folklore into literature, art, and science, let alone into film and music.

5 The Messages of Cats

With the aid of human words and images, cats – kitties – pussies – tabbies – moggies – historically, grimalkins – can and do convey many messages. They star in countless proverbs; and in colloquial idioms.[67] To name but a few: one who falls unduly silent has let the 'cat get his or her tongue'. Accidentally revealing a secret is 'letting cat out of the bag'. Someone attempting the impossible has 'not a cat in hell's chance'. A heavy downpour means that 'it's raining cats and dogs'.[68] In a confined space, there is 'no room to swing a cat'. Something extremely ridiculous is 'enough to make a cat laugh'. A timorous person is a 'scaredy cat'. A slavish imitator is a 'copy-cat'. And a self-satisfied plutocrat, who flaunts great wealth from suspect sources, is undeniably a 'fat cat'.

Particularly celebrated among feline messages in the eighteenth century was that penned by Thomas Gray. His friend Horace Walpole reported that a treasured pet had met an untimely end. Thereupon Gray's *Ode on the Death of*

63 Susan Chitty: The Singular Person Called Lear. The Extraordinary Life of the Author of the Owl and the Pussycat. Stroud 2007.

64 Langford Reed: The Life of Lewis Carroll. London 1932; Lewis Carroll: The Annotated Alice. Ed. by Martin Gardner. Harmondsworth 1974.

65 For full amplification, see Will Brooker: Alice's Adventures. Lewis Carroll and Alice in Popular Culture. New York 2004.

66 For further reflections upon scientific/feline imagery, see John Gribbin: In Search of Schrödinger's Cat. The Startling World of Quantum Physics Explained. Hounslow 1984; Paul Doherty and Don Rathjen: The Cheshire Cat and Other Eye-Popping Experiments on How We See the World. New York 1995.

67 For browsing on this theme, see Robert A. Nowlan and Gwendolyn L. Nowlan (ed.): A Dictionary of Quotations and Proverbs about Cats and Dogs. London 2000.

68 People in many cultures use many graphic expressions for heavy rainfall.

FIGURES 12.5–12.6 John Tenniel's justly classic depictions of Lewis Carroll's 'Cheshire Cat', from *Alice's Adventures in Wonderland* (1865), as the Cat appears and slowly disappears, leaving nothing but its grin

a Favourite Cat, Drowned in a Tub of Goldfishes (1748) re-imagined the fate of the once-purring Selina, "demurest of the tabby kind". Her tail twitches as she gazes eagerly upon two lustrous goldfish in their bowl.

> With many an ardent wish
> She stretched, in vain, to reach the prize –
> What female heart can gold despise?
> What Cat's averse to Fish?

But alas the animal plunges in and drowns. Her mews for help go unheeded. The poet then offers a sage warning to all covetous females.

> Not all that tempts your wandering eyes
> And heedless hearts, is lawful prize,
> Nor all that glisters, gold« .[69]

The poem is charmingly wrought. And the message apt enough to warn all humans.

Counteracting such sagacious caution, however, another message celebrated feline adventurousness. Cats were prepared to move when human homes moved. Accordingly, thousands of feline pest controllers travelled the high seas. They dined on the mice and rats which infested the wooden ships of the commercial fleet and the Royal Navy.[70] (Keeping pets on board was banned by the British Navy only in 1975). Often these sailing cats were adopted by the crews as informal mascots. In the years 1740–44, one is known to have circumnavigated the globe with Admiral George Anson. And this well-travelled cat may be commemorated by the lofty Cat Monument (designed 1749) at Shugborough Park in Staffordshire. It was erected by the Admiral's older brother, Thomas Anson. He too was a felinophile. So the Monument may instead be celebrating Anson's rare Persian cat.[71] Either way, a lordly feline with luxuriant whiskers crouches atop a huge Grecian urn, at the ready.

---

69 For all quotations about the demure Selina, see Thomas Gray: On the Death of a Favourite Cat, Drowned in a Tub of Goldfishes [1748]. In: Francis Turner Palgrave (ed.): The Golden Treasury of the best Songs and lyrical Poems in the English Language [1861; 1959]. Basingstoke 2000, p. 138–139.

70 For context, see Val Lewis: Ship's Cats in War and Peace. Shepperton-on-Thames 2001.

71 For the Cat Monument (designed 1749; erected c. 1770) NGR: SJ9932022722; online: https://www.nationaltrust.org.uk/shugborough-estate/features/parkland-buildings-monuments-and-follies [09.12.2020].

Also famed as a circumnavigator, this time of Australia, was a black cat named Trim, after the devoted Corporal Trim in Sterne's novel *Tristram Shandy* (1759). This feisty feline sailed with the explorer Matthew Flinders in 1801/02. At one point, the cat was washed overboard. But it swam to a rope and clambered up to safety, winning the captain's admiration. Flinders later wrote a glowing personal tribute to Trim, calling him "one of the finest animals I ever saw" and "the best and most illustrious of his race".[72] Today the two of them are commemorated in several pleasant statues at different locations in Australia and England.

Trim is thus the real-life archetype of all travelling cats, who, in fiction, now travel in spacecraft as well.[73] With humans, they look novelty calmly in the eye. A children's nursery rhyme (1805) chanted the reply of a Pussycat, who, when challenged, explains that she has been: "To London to visit the Queen". No imagined social nerves on her part. Indeed, traditional lore had long insisted that "a cat may look at a king". Anyone, whether low or high, could look monarchy squarely in the face. Musing on this dictum in 1713, a conservative cultural commentator remarked somewhat sourly that it was: "a true-blue Protestant Proverb".[74] It signalled a proud and potentially political egalitarianism.

Above and beyond all other feline messages, was, however, their link with literary creativity and intellectual enquiry. No disrespect to other indoor pets. Or to those magnificent outdoor companions: dogs[75] and horses.[76] But, as already noted, the feline admixture of caution and curiosity made them potent triggers to innovative thought. There was an accumulating 'cat strand' of ideas – celebrating the amiable, the wily and the enigmatic feline. And it has multiplied

72 See The Finders Papers FLI/11 for Matthew Flinders' *Biographical Tribute to his Cat Trim* (1809); online: https://flinders.rmg.co.uk/DisplayDocument2410.html [04.01.2021], p. 1; later published as Matthew Flinders: Trim: Being the True Story of a Brave Seafaring Cat. Pymble/NSW 1997. The praise of Trim as "the best and most illustrious of his race" is engraved on the plaque dedicated to Trim at the Mitchell Library, Sydney.

73 An example is the cat Spot in the sci-fi TV series Star Trek (1966–).

74 For both the proverbial quotation and his commentary, see Oswald Dykes: English Proverbs, with Moral Reflections. London 1713, p. 41.

75 Frank Jackson (ed.): Faithful Friends: Dogs in Life and Literature. London 1997; Keridiana W. Chez: Victorian Dogs, Victorian Men. Affect and Animals in Nineteenth-Century Literature and Culture. Columbus/OH 2017; Michelle Marie Taylor: From Sentiment to Sagacity to Subjectivity: Dogs and Genre in Nineteenth-Century British Literature. [unpublished PhD thesis, Iowa University 2018]; online: https://doi.org/10.17077/etd.aa4yzdvq [03.12.2020].

76 See Karen L. Raber and Treva J. Tucker (ed.): The Culture of the Horse. Status, Discipline, and Identity in the Early Modern World. New York 2005, passim.

FIGURE 12.7 Bronze statue of Trim, the feisty black cat which circumnavigated Australia with explorer Matthew Flinders in 1801/02: located outside Mitchell Library, 173 Macquarie St, Sydney, Australia

in later eras, in many cultures,[77] and in many media, including films and musicals.[78] In 1997 the French shock-master Jacques Derrida joined the philosophic debates by meditating on his cat's gaze upon his own naked body.[79] In turn, these varied messages strike varied chords with the felinophile public, encouraging further creative output and debates. So, while the close relationship of cats and humans goes back for millennia, it thrives upon imaginative updating.

77 A great example from Japan is Natsume Sōseki's humorous satire *I am a Cat* (1906).

78 Thomas Stearns Eliot's cat-poems from the 1930s appeared as *Old Possum's Book of Practical Cats* (1939), later adapted into Andrew Lloyd Webber's hit musical *Cats!* (premiered 1981).

79 See Jacques Derrida: The Animal That Therefore I am. Transl. by David Wills. Ed. by Marie-Luise Mallet, Ashland/OH 2008 [French: *L'animal que donc je suis* (1997)]; and discussion in Donna J. Haraway: When Species Meet. Minneapolis 2008, p. 19–20; see also John Gray: Feline Philosophy: Cats and the Meaning of Life. London 2020.

The process is salutary. Humans are speaking animals, sharing a world with countless living creatures, whose kinship should not be forgotten.

Christopher Smart provides a final thought. He lauded what he saw as the God-given spiritual fire of his cat Jeoffry. And Smart described the animal's potency in modish terms. Mid-eighteenth-century scientists were striving to tap a new source of power.[80] For Smart, that stimulus was provided by the dynamic effect of a cat: "For by stroking of him [Jeoffry], I have found out electricity".[81]

Acknowledgements

With warm thanks for insightful comments to colleagues in the Virtual Potsdam Conference, 5 August 2020, as well as to Tony Belton, James Dixon, and especially my fellow editors, who devised this project, Stefanie Stockhorst and Jürgen Overhoff.

80 For scientific and cultural context, see Patricia Fara: An Entertainment for Angels. Electricity in the Enlightenment. London 2017.

81 Smart, Jubilate Agno, p. 44 and p. 124.

Index of Names / Index des noms

Index of Non-Human Animals / Index des animaux non-humains

www.ingramcontent.com/pod-product-compliance
Lightning Source LLC
Chambersburg PA
CBHW071322230926
51849CB00002B/290
* 9 7 8 9 0 0 4 4 4 8 7 2 8 *